INUIT FOLK-TALES

Collected By
Knud Rasmussen

with an introduction by
Birgit Kleist Pedersen and Jette Rygaard

Afterword and Translation by
W. Worster

Distributed by University Press of New England
Hanover and London

Reprint of 1921 edition titled *Eskimo Folk-tales.*

Distributed by University Press of New England,
One Court Street, Lebanon, NH 03766
www.upne.com
Funded in part by Greenland Home Rule Government, Nuuk
Cover image courtesy NASA, Visible Images.
ISBN 978-0-9821703-1-1

IPI Press

Post Office Box 212
Hanover, New Hampshire 03755 USA
•
McGill University, Arts Building
853 Sherbrooke St. West
Montreal, PQ, H3A-2T6 Canada

CONTENTS

INTRODUCTION

Knud Johan Victor Rasmussen (1879-1933), known best in Greenland as "Kununnguaq", was recognized internationally as an Arctic researcher, explorer, writer, and collector of the Greenlandic oral tradition of the tale. As a native speaker, born to a Greenlandic mother and Danish father, he was raised in Jakobshavn/Ilulissat until 1891. Then he travelled to Denmark for further education. When opportunity arose later in his life, it was this unique background that made him the perfect interpreter of native customs and gave him the familiarity needed to collect the tales of his homeland and to communicate them in written form for both the Greenlandic people and their kinsmen to the West.

When Rasmussen left Greenland to pursue an education in Denmark, he contemplated using his linguistic knowledge to study comparative linguistics. Instead, he took singing lessons to pursue a career as an opera singer and an actor. He also toyed with the idea of becoming a writer.

After a short period as a journalist, the turning point of his life came when he encountered the polar explorers and writers Mylius Erichsen and Fritdtjof Nansen. They were involved in organizing an expedition to Greenland – The Literary Expedition. Rasmussen had no doubts about the importance of these meetings or his desire to join the expedition. He wrote to his father:

The Literary Greenlandic Expedition has attracted the most attention ... Nansen and I 'found each other' yesterday after a constant dialogue from 3-7. He has great hopes for the good profit of our travel. He thinks that I - through my knowledge of the Greenlandic language have - 'a future as ethnographer' – if I will make use of the – 'very rare qualifications' – I hold. [...] If it interests you – dear Father – Your son is going to a party tonight at Professor Sars, where I am going to meet the scientific Norths. My friend Nansen will be there too. [...] Now I don't dare to write any more, lest, you will think that I am becoming conceited. I send you a picture of the palace of Nansen that was in the paper 'The Course of the World' yesterday. When I become world-famous in some years – and have become well off from my fame like N – then I will build a palace like that ...' (Translated from Danish, JR)

Rasmussen's most important publications of tales, "Myter og Sagn fra Grønland", vol. 1-3 (1921-25), are derived from the material he recorded in East Greenland, West Greenland and from Cape York during the Literary Expedition between 1902 and 1904. Rasmussen travelled along the whole West coast, collecting Greenlandic folklore and local traditions. He is still remembered as a 'festive spirit' by people who have met him in person. The advantage of having Greenlandic blood in his veins gave him access, as well as status, as a collector of manuscripts and as a recorder of oral tales (paid with small fees). He explained his method of obtaining tales in the following manner:

. . . first I let the teller give me his tales from start to finish without interruption. Thereafter I wrote down the legend, as I made him repeat it sentence by sentence. Through daily work with the same material, you acquire such an immense training in perceiving and remembering, that you, in the end yourself, learn to know the tale simultaneously while listening to it. Should it occur that the teller becomes tired of the lengthy transcription and therefore tries to shorten or skimp the original rendering, you will immediately be able to intervene and guide him back to the original form. (Translated from Danish, BKP)

Upon inspection of the transcriptions of Rasmussen at The Royal National Library's Archives in Copenhagen (Det Kgl. Biblioteks Rigsarkiv), it appears that Rasmussen does makes some literary changes in voice when the tales are collected and published. As the Danish eskimologist, Dr. Kirsten Thisted points out in her PhD dissertation on the Greenlandic Oral Tradition (Som Perler på en Snor

(Like Beads on a String), Copenhagen, 1993), it is conspicuous that all Rasmussen's published tales from every part of Greenland are marked with the same style of telling, in spite of the geographic and cultural distances. The reason, Thisted assumes, is Rasmussen's own voice:

> *This is due to the circumstance that Rasmussen is not presenting the individual teller, on the contrary it is (..) 'the voice of the people'. ...Rasmussen tries to reach the 'authentic' foundation, the people as they originally used to be; before the colonization and Christianizing, in its purity.* (Translated from Danish, BKP)

It is this devotion to ethnography, as well as his sensitivity and curiosity that made Rasmussen the scientific and literary giant that he still is today. These tales were recorded along with customs and artefacts of Greenland, by a man interested in both the preservation and growth of his country.

THE TWO FRIENDS WHO SET OFF TO TRAVEL ROUND THE WORLD

ONCE there were two men who desired to travel round the world, that they might tell others what was the manner of it.

This was in the days when men were still many on the earth, and there were people in all the lands. Now we grow fewer and fewer. Evil and sickness have come upon men. See how I, who tell this story, drag my life along, unable to stand upon my feet.

The two men who were setting out had each newly taken a wife, and had as yet no children. They made themselves cups of musk-ox horn, each making a cup for himself from one side of the same beasts head. And they set out, each going away from the other, that they might go by different ways and meet again some day. They travelled with sledges, and chose land to stay and live upon each summer.

It took them a long time to get round the world; they had children, and they grew old, and then their children also grew old, until at last the parents were so old that they could not walk, but the children led them.

And at last one day, they met and of their drinking horns there was but the handle left, so many times had they drunk water by the way, scraping the horn against the ground as they filled them. "The world is great indeed" they said when they met.

They had been young at their starting, and now they were old men, led by their children.

Truly the world is great.

THE COMING OF MEN, A LONG, LONG WHILE AGO

OUR forefathers have told us much of the coming of earth, and of men, and it was a long, long while ago. Those who lived long before our day, they did not know how to store their words in little black marks, as you do; they could only tell stories. And they told of many things, and therefore we are not without knowledge of these things, which we have heard told many and many a time, since we were little children. Old women do not waste their words idly, and we believe what they say. Old age does not lie.

A long, long time ago, when the earth was to be made, it fell down from the sky. Earth, hills and stones, all fell down from the sky, and thus the earth was made.

And then, when the earth was made, came men.

It is said that they came forth out of the earth. Little children came out of the earth. They came forth from among the willow bushes, all covered with willow leaves. And there they lay among the little bushes: lay and kicked, for they could not even crawl. And they got their food from the earth.

Then there is something about a man and a woman, but what of them? It is not clearly known. When did they find each other, and when had they grown up? I do not know. But the woman sewed, and made children's clothes, and wandered forth. And she found little children, and dressed them in the clothes, and brought them home.

And in this way men grew to be many.

And being now so many, they desired to have dogs. So a man went out with a dog leash in his hand, and began to stamp on the ground, crying "Hok hok hok!" Then the dogs came hurrying out from the hummocks, and shook themselves violently, for their coats were full of sand. Thus men found dogs.

But then children began to be born, and men grew to be very many on the earth. They knew nothing of death in those days, a long, long time ago, and grew to be very old. At last they could not walk, but went blind, and could not lie down.

Neither did they know the sun, but lived in the dark. No day ever dawned. Only inside their houses was there ever light, and they burned water in their lamps, for in those days water would burn.

But these men who did not know how to die, they grew to be too many, and crowded the earth. And then there came a mighty flood from the sea. Many were drowned, and men grew fewer. We can still see marks of that great flood, on the high hill-tops, where mussel shells may often be found.

And now that men had begun to be fewer, two old women began to speak thus:

“Better to be without day, if thus we may be without death," said the one. "No; let us have both light and death," said the other.

And when the old woman had spoken these words, it was as she had wished. Light came, and death.

It is said, that when the first man died, others covered up the body with stones. But the body came back again, not knowing rightly how to die. It stuck out its head from the bench, and tried to get up. But an old woman thrust it back, and said:

"We have much to carry, and our sledges are small."

For they were about to set out on a hunting journey. And so the dead one was forced to go back to the mound of stones.

And now, after men had got light on their earth, they were able to go on journeys, and to hunt, and no longer needed to eat of the earth. And with death came also the sun, moon and stars.

For when men die, they go up into the sky and become brightly shining things there.

NUKUNGUASIK, WHO ESCAPED FROM THE TUPILAK*

NUKUNGUASIK, it is said, had land in a place with many brothers. When the brothers made a catch, they gave him meat for the pot; he himself had no wife.

One day he rowed northward in his kayak, and suddenly he took it into his head to row over to a big island which he had never visited before, and now wished to see. He landed, and went up to look at the land, and it was very beautiful there.

**Tupilak: a monster created by one having magic powers, who uses it to wreak vengeance on an enemy.*

And here he came upon the middle one of many brothers, busy with something or other down in a hollow, and whispering all the time. So he crawled stealthily towards him, and when he had come closer, he heard him whispering these words:

"You are to bite Nukunguasik to death; you are to bite Nukunguasik to death."

And then it was clear that he was making a Tupilak, and stood there now telling it what to do. But suddenly Nukunguasik slapped him on the side and said: "But where is this Nukunguasik?"

And the man was so frightened at this that he fell down dead.

And then Nukunguasik saw that the man had been letting the Tupilak sniff at his body. And the Tupilak was now alive, and lay there sniffing. But Nukunguasik, being afraid of the Tupilak, went away without trying to harm it.

Now he rowed home, and there the many brothers

were waiting in vain for the middle one to return. At last the day dawned, and still he had not come. And daylight came, and then as they were preparing to go out in search of him, the eldest of them said to Nukunguasik:

"Nukunguasik, come with us; we must search for him."

And so Nukunguasik went with them, but as they found nothing, he said:

"Would it not be well to go and make search over on that island, where no one ever goes?"

And having gone on to the island, Nukunguasik said:

"Now you can go and look on the southern side."

When the brothers reached the place, he heard them cry out, and the eldest said:

"O wretched one! Why did you ever meddle with

such a thing as this!"

And they could be heard weeping all together about the dead man.

And now Nukunguasik went up to them, and there lay the Tupilak, still alive, and nibbling at the body of the dead man. But the brothers buried him there, making a mound of stones above him. And then they went home.

Nukunguasik lived there as the oldest in the place, and died at last after many years.

Here I end this story: I know no more.

QUJAVARSSUK

A STRONG man had land at Ikerssuaq. The only other one there was an old man, one who lived on nothing but devil fish; when the strong man had caught more than he needed, the old man had always plenty of meat, which was given him in exchange for his fish.

The strong one, men say, he who never failed to catch seal when he went out hunting, became silent as time went on, and then very silent. And this no doubt was because he could get no children.

The old one was a wizard, and one day the strong one came to him and said:

"To-morrow, when my wife comes down to the shore close by where you are fishing, go to her. For this I will give you some thing of my catch each day."

And this no doubt was because he wanted his wife to have a child, for he wished greatly to have a child, and could not bring it about.

The old man did not forget those words which were said to him.

And to his wife also, the strong one said:

"To-morrow, when the old one is out fishing, go you down finely dressed, to the shore close by."

And she did it as he had said. When they had slept and again awakened, she watched to see when the old one went out. And when he rowed away, she put on her finest clothes and followed after him along the shore. When she came in sight of him, he lay out there fishing. Then eagerly she stood up on the shore, and looked out towards him. And now he looked at her, and then again out over the sea, and this went on for a long time. She stood there a long time in vain, looking out towards him, but he would not come in to where she was, and therefore she went home. As soon as she had come home, her husband rowed up to the old one, and asked:

"Did you not go to my wife to-day?"

The old one said:

"No."

And again the strong one said a second time:

"Then do not fail to go to her to-morrow."

But when the old one came home, he could not forget the strong man s words. In the evening, the strong one said that same thing again to his wife, and a second time told her to go to the old one.

They slept, and awakened, and the strong man went out hunting as was his wont. Then his wife waited only until the old one had gone out, and as soon as he was gone, she put on her finest clothes and followed after. When she came in sight of the water, the old one was sitting there in his boat as on the other days, and fishing. Now the old one turned his head and saw her, and he could see that she was even more finely dressed than on the day before. And now a great desire of her came over him, and he made up his mind to row in to where she was. He came in to the land, and stepped out of his kayak and went up to her. And now he went to her this time.

Then he rowed out again, but he caught scarcely any fish that day.

When only a little time had gone, the strong man came rowing out to him and said:

"Now perhaps you have again failed to go to my wife?"

When these words were spoken, the old one turned his head away, and said:

"To-day I have not failed to be with her."

When the strong one heard this, he took one of the seals he had caught, and gave it to the old man, and said:

"Take this; it is yours."

And in this way he acted towards him from that time. The old one came home that day dragging a seal behind him. And this he could often do thereafter.

When the strong one came home, he said to his wife:

"When I go out to-morrow in my kayak, it is not to hunt seal; therefore watch carefully for my return when the sun is in the west."

Next day he went out in his kayak, and when the sun was in the west, his wife went often and often to look out. And once when she went thus, she saw that he had come, and from that moment she was no longer sleepy.

As the strong one came nearer and nearer to land, he paddled more and more strongly.

Now his wife went down to that place where he was about to land, and turned and sat down with her back to the sea. The man unfastened his hunting fur from the ring of his kayak, and put his hand into the back of the kayak, and took out a sea serpent, and struck his wife on the back. At this she felt very cold, and her skin smarted. Then she stood up and went home. But her husband said no word to her. Then they slept, and awakened, and then the old one came to them and said:

"Now you must search for the carrion of a cormorant, with only the skeleton remaining, for your wife is with child."

And the strong one went out eagerly to search for this.

One day, paddling southward in his kayak, as was his custom, he started to search all the little bird cliffs. And coming to the foot of one of them, he saw that which he so greatly wished to see; the carrion of a big cormorant, which had now become a skeleton. It lay there quite easy to see. But there was no way of coming to the place where it was, not from above nor from below, nor from the side. Yet he would try. He tied his hunting line fast to the cross thongs on his kayak, and thrust his hand into a small crack a little way up the cliff. And now he tried to climb up there with his hands alone. And at last he got that skeleton, and came down in the same way back to his kayak, and got into it, and rowed away northward to his home. And almost before he had reached land, the old one came to him, and the cormorant skeleton was taken out of the kayak. Now the old one trembled all over with surprise. And he took the skeleton, and put it away, and said:

"Now you must search for a soft stone, which has never felt the sun, a stone good to make a lamp of."

And the strong man began to search for such a stone.

Once when he was on this search, he came to a cliff, which stood in such a place that it never felt the sun, and here he found a fine lamp stone. And he brought it home, and the old one took it and put it away.

A few days passed, and then the strong one's wife began to feel the birth-pangs, and the old one went in there at once with his own wife. Then she bore a son, and when he was born, the strong man said to the old one:

"This is your child; name him after some dead one." *

"Let him be named after him who died of hunger in the north, at Amerdloq."

This the old one said. And then he said:
"His name shall be Qujavarssuk!"

**According to custom. It is believed that the qualities of the dead are thus transferred to the living namesake.*

And in this way the old one gave him that name.

Now Qujavarssuk grew up, and when he was grown big enough, the strong man said to the old one:

"Make a kayak for him."

Now the old one made him a kayak, and the kayak was finished. And when it was finished, he took it by the nose and thrust him out into the water to try it, but without loosing his hold. And when he did this, there came one little seal up out of the water, and others also. This was a sign that he should be a strong man, a chief, when the seals came to him so. When he drew him out of the water, they all went down again, and not a seal remained.

Now the old one began to make hunting things. When they were finished, and there was nothing more to be done in making them, and he thought the boy was of a good age to begin going out to hunt seal, he said to the strong one:

"Now row out with him, for he must go seal hunting."

Then he rowed out with him, and when they had come so far out that they could not see the bottom, he said:

"Take the harpoon point with its line, and fix it on the shaft."

They had just made things ready for their hunting and rowed on farther, when they came to a flock of black seal.

The strong one said to him:

"Now row straight at them."

And then he rowed straight at them, and he lifted his harpoon and he threw it and he struck. And this he did every day in the same manner, and made a catch each time he went out in his kayak.

Then some people who had made a wintering place in the south heard, in a time of hunger, of Quja-varssuk, the strong man who never suffered want. And when they heard this, they began to come and visit the place where he had land. In this way there came once a man who was called Tugto, and his wife. And while they were there they were both great wizards the man and his wife began to quarrel, and so the wife ran away to live alone in the hills. And now the man could not bring back his wife, for he was not so great a wizard as she. And when the people who had come to visit the place went away, he could do nothing but stay there.

One day when he was out hunting seal at Ikers-suaq, he saw a big black seal which came up from the bottom with a red fish in its mouth.

Now he took bearings by the cliffs of the place where the seal went down, and after that time, when he was out in his kayak, he took up all the bird wings that he saw, and fastened all the pinion feathers together.

Tugto was a big man, yet he had taken up so much of this that it was a hard matter for him to carry it when he took it on his back, and then he thought it must be enough for that depth of water.

At last the ice lay firm, and when the ice lay firm, he began to make things ready to go out and fish. One morning he woke, and went away over land. He came to a lake, and walked over it, and came again on to the land. And thus he came to the place where lay that water he was going to fish, and he went out on the ice while it was still morning. Then he cut a great hole in the ice, and just as he cast out the weight on his line, the sun came up. It came quite out, and went across the sky, all in the time he was letting out his line. And not until the sun had gone half through the day did the weight reach the bottom. Then he hauled up the line a little way, and almost before it was still, he felt a pull. And he hauled it up, and it was a mighty sea perch. This he killed, but did not let down his line a second time, for in that way it would become evening. He cut a hole in the lower jaw of the fish, and put in a cord to carry it with. And when he took it on his head, it was so long that the tail struck against his heel.

Then in this manner he walked away, and came to land. When he came to the big lake he had walked over in the morning, he went out on it. But when he had come half the way over, the ice began to make a noise, and when he looked round, it seemed to him that the noise in the ice was following him from behind.

Now he went away running, but as he ran he fainted suddenly away, and lay a long time so. When he woke again, he was lying down. He thought a little, and then he remembered. "I am running away!" And then he got up and turned round, but could not find a break in the ice anywhere. But he could feel in himself that he had now become a much greater wizard than before.

He went on farther, and chose his way up over a little hilly slope, and when he could see clearly ahead, he perceived a mighty beast.

It was one of those monsters which men saw in the

old far-off times, quite covered with bird-skins. And it was so big that not a twitch of life could be seen in it. He was afraid now, and turned round, until he could no longer see it. Then he left that way, and came out into another place, where he saw another looking just the same. He now went back again in such a manner that it could not find him, but then he remembered that a wizard can win power to vanish away, even to vanish into the ground, if he can pull to pieces the skin of such a monster.

When his thoughts had begun to work upon this, he threw away his burden and went towards it and began to wrestle with it. And it was not a long time before he began to tear its covering in pieces; the flesh on it was not bigger than a thumb. Then he went away from it, and took up his burden again on his head, and went wandering on. When he was again going along homewards, he felt in himself that he had become a great wizard, and he could see the door openings of all the villages in that countryside quite close together.

And when he came home, he caused these words to be said:

"Let the people come and hear."

And now many people came hurrying into the house. And he began calling up spirits. And in this calling he raised himself up and flew away towards his wife.

And when he came near her in his spirit flight, and hovered above her, she was sitting sewing. He went straight down through the roof, and when she tried to escape through the floor he did likewise, and reached her in the earth. After this, she was very willing when he tried to take her home with him, and he took her home with him, and now he had his wife again, and those two people lived together until they were very old.

One winter, the frost came, and was very hard and the sea was frozen solid; and only a little opening was left, far out over the ice. And hither Qujavarssuk was forced to carry his kayak each day, out to the open water, but each day he caught two seals, as was his custom.

And then, as often happens in time of dearth, there came many poor people wandering over the ice, from the south, wishing to get some good thing of all that Qujavarssuk caught. Once there came also two old men, and they were his mother's kinsmen. They came on a visit. And when they came, his mother said to them:

"Now you have come before I have got anything cooked. It is true that I have something from the cooking of yesterday; eat that if you will, while I cook something now." Then she set before them the kidney part of a black seal, with its own blubber as dripping. Now one of the two old men began eating, and went on eagerly, dipping the meat in the dripping. But the other stopped eating very soon.

Then Qujavarssuk came home, as was his custom, with two seals, and said to his mother:

"Take the breast part and boil it quickly."

For this was the best part of the seal. And she boiled it, and it was done in a moment. And then she set it on a dish and brought it to those two.

"Here, eat."

And now at last the one of them began really to eat, but the other took a piece of the shoulder. When Qujavarssuk saw this, he said:

"You should not begin to eat from the wrong side."

And when he had said that, he said again:

"If you eat from that side, then my catching of the seals will cease." But the old man became very angry in his mind at this order.

Next morning, when they were about to set off again southward, Qujavarssuk's mother gave them as much meat as they could carry. They went home southward, over the ice, but when they had gone a little way, they were forced to stop, because their burden was so heavy. And when they had rested a little, they went on again. When they had come near to their village, one said to the other:

"Has there not wakened a thought in your mind? I am very angry with Qujavarssuk. Yesterday, when we came there, they gave us only a kidney piece in welcome, and that is meat I do not like at all."

"Hum," said the other. "I thought it was all very good. It was fine tender meat for my teeth."

At these words, the other began again to speak:

"Now that my anger has awakened, I will make a Tupilak for that miserable Qujavarssuk."

But the other said to him:

"Why will you do such a thing? Look; their gifts are so many that we must carry the load upon our heads."

But that comrade would not change his purpose, not for all the trying of the other to turn him from it. And at last the other ceased to speak of it.

Now as the cold grew stronger, that opening in the ice became smaller and smaller, at the place where Qujavarssuk was used to go with his kayak. One day, when he came down to it, there was but just room for his kayak to go in, and if now a seal should rise, it could not fail to strike the kayak. Yet he got into the kayak, and at the time when he was fixing the head on his harpoon, he saw a black seal coming up from below. But seeing that it must touch both the ice and the kayak, it went down again without coming right to the surface. Then Qujavarssuk went up again and went home, and that was the first time he went home without having made a catch, in all the time he had been a hunter.

When he had come home, he sat himself down behind his mother s lamp, sitting on the bedplace, so that only his feet hung down over the floor. He was so troubled that he would not eat. And later in the evening, he said to his mother:

"Take meat to Tugto and his wife, and ask one of them to magic away the ice."

His mother went out and cut the meat of a black seal across at the middle. Then she brought the tail half, and half the blubber of a seal, up to Tugto and his wife. She came to the entrance, but it was covered with snow, so that it looked like a fox hole. At first, she dropped that which she was carrying in through the passage way. And it was this which Tugto and his wife first saw; the half of a black seal's meat and half of its blubber cut across. And when she came in, she said:

"It is my errand now to ask if one of you can magic away the ice."

When these words were heard, Tugto said to his wife: "In this time of hunger we cannot send away meat that is given. You must magic away the ice."

And she set about to do his bidding. To Quja-varssuk's mother she said:

"Tell all the people who can come here to come here and listen!"

And then she began eagerly going in to the dwellings, to say that all who could come should come in and listen to the magic. When all had come in, she put out the lamp, and began to call on her helping spirits. Then suddenly she said:

"Two flames have appeared in the west!"

And now she was standing up in the passage way, and let them come to her, and when they came forward, they were a bear and a walrus. The bear blew her in under the bedplace, but when it drew in its breath again, she came out from under the bedplace and stopped at the passage way. In this manner it went on for a long time. But now she made ready to go out, and said then to the listeners:

"All through this night none may yawn or wink an eye." And then she went out.

At the same moment when she went out, the bear took her in its teeth and flung her out over the ice. Hardly had she fallen on the ice again, when the walrus thrust its tusks into her and flung her out across the ice, but the bear ran along after her, keeping beneath her as she flew through the air. Each time she fell on the ice, the walrus thrust its tusks into her again. It seemed as if the outermost islands suddenly went to the bottom of the sea, so quickly did she move outwards. They were now almost out of sight, and not until they could no longer see the land did the walrus and the bear leave her. Then she could begin again to go towards the land.

When at last she could see the cliffs, it seemed as if there were clouds above them, because of the driving snow. At last the wind came down, and the ice began at once to break up. Now she looked round on all sides, and caught sight of an iceberg which was frozen fast. And towards this she let herself drift. Hardly had she come up on to the iceberg, when the ice all went to pieces, and now there was no way for her to save herself. But at the same moment she heard someone beside her say:

"Let me take you in my kayak." And when she looked round, she saw a man in a very narrow kayak. And he said a second time:

"Come and let me take you in my kayak. If you will not do this, then you will never taste the good things Qujavarssuk has paid you."

Now the sea was very rough, and yet she made ready to go. When a wave lifted the kayak, she sprang down into it. But as she dropped down, the kayak was nearly upset. Then, as she tried to move over to the other side of it, she again moved too far, and then he said:

"Place yourself properly in the middle of the kayak."

And when she had done so, he tried to row, for it

was his purpose to take her with him in his kayak, although the sea was very rough. Then he rowed out with her. And when he had come a little way out, he sighted land, but when they came near, there was no place at all where they could come up on shore, and at the moment when the wave took them, he said:

"Now try to jump ashore."

And when he said this, she sprang ashore. When she now stood on land, she turned round and saw that the kayak was lost to sight in a great wave. And it was never seen again. She turned and went away. But as she went on, she felt a mighty thirst. She came to a place where water was oozing through the snow. She went there, and when she reached it, and was about to lay herself down to drink, a voice came suddenly and said:

"Do not drink it; for if you do, you will never taste the good things Qujavarssuk has paid you."

When she heard this she went forward again. On

her way she came to a house. On the top of the house lay a great dog, and it was terrible to see. When she began to go past it, it looked as if it would bite her. But at last she came past it.

In the passage way of the house there was a great river flowing, and the only place where she could tread was narrow as the back of a knife. And the passage way itself was so wide that she could not hold fast by the walls.

So she walked along, poising carefully, using her little fingers as wings. But when she came to the inner door, the step was so high, that she could not come over it quickly. Inside the house, she saw an old woman lying face downwards on the bedplace. And as soon as she had come in, the old woman began to abuse her. And she was about to answer those bad words, when the old woman sprang out on to the floor to fight with her. And now they two fought furiously together. They fought for a long time, and little by little the old woman grew tired. And when she was so tired that she could not get up, the other saw that her hair hung loose and was full of dirt. And now Tugto's wife began cleaning her as well as she could. When this was done, she put up her hair in its knot. The old woman had not spoken, but now she said:

"You are a dear little thing, you that have come in here. It is long since I was so nicely cleaned. Not since little Atakana from Sardloq cleaned me have I ever been cleaned at all. I have nothing to give you in return. Move my lamp away."

And when she did so, there was a noise like the moving of wings. When she turned to look, she saw a host of birds flying in through the passage way. For a long time birds flew in, without stopping. But then the woman said:

"Now it is enough." And she put the lamp straight. And when that was done, the other said again:

"Will you not put it a little to the other side?"

And she moved it so. And then she saw some men with long hair flying towards the passage way.

When she looked closer, she saw that it was a host of black seal. And when very many of them had come in this manner, she said:

"Now it is enough." And she put the lamp in its place. Then the old woman looked over towards her, and said:

"When you come home, tell them that they must never more face towards the sea when they empty their dirty vessels, for when they do so, it all goes over me."

When at last the woman came out again, the big dog wagged his tail kindly at her.

It was still night when Tugto's wife came home, and when she came in, none of them had yet yawned or winked an eye. When she lit the lamp, her face was fearfully scratched, and she told them this:

"You must not think that the ice will break up at once; it will not break up until these sores are healed."

After a long time they began to heal slowly, and sometimes it might happen that one or another cried in mockingly through the window:

"Now surely it is time the ice broke up and went out to sea, for that which was to be done is surely done."

But at last her sores were healed. And one day a black cloud came up in the south. Later in the evening, there was a mighty noise of the wind, and the storm did not abate until it was growing light in the morning. When it was quite light, and the people came out, the sea was open and blue. A great number of birds were flying above the water, and there were hosts of black seal everywhere. The kayaks were made ready at once, and when they began to make them ready, Tugto's wife said:

"No one must hunt them yet; until five days are gone no one may hunt them."

But before those days were gone, one of the young

men went out nevertheless to hunt. He tried with great efforts, but caught nothing after all. Not until those days were gone did the witch-wife say:

"Now you may hunt them."

And now the men went out to sea to hunt the birds. And not until they could bear no more on their kayaks did they row home again. But then all those men had to give up their whole catch to Tugto's house. Not until the second hunting were they permitted to keep any for themselves.

Next day they went out to hunt for seal. They harpooned many, but these also were given to Tugto and his wife. Of these also they kept nothing for themselves until the second hunting.

Now when the ice was gone, then that old man we have told about before, he put life into the Tupilak, and said to it then:

"Go out now, and eat up Qujavarssuk."

The Tupilak paddled out after him, but Qujavarssuk had already reached the shore, and was about to carry up his kayak on to the land, with a catch of two seals. Now the Tupilak had no fear but that next day, when he went out, it would be easy to catch and eat him. And therefore, when it was no later than dawn, it was waiting outside his house. When Qujavarssuk awoke, he got up and went down to his kayak, and began to make ready for hunting. He put on his long fur coat, and went down and put the kayak in the water. He lifted one leg and stepped into the kayak, and this the Tupilak saw, but when he lifted the other leg to step in with that, he disappeared entirely from its sight. And all through the day it looked for him in vain. At last it swam in towards land, but by that time he had already reached home, and drawn the kayak on shore to carry it up. He had a catch of two seal, and there lay the Tupilak staring after him.

When it was evening, Qujavarssuk went to rest. He slept, and awoke, and got up and made things ready to go out. And at this time the Tupilak was waiting with a great desire for the moment when he should put off from land. But when he put on his hunting coat ready to row out, the Tupilak thought:

"Now we shall see if he disappears again."

And just as he was getting into his kayak, he disappeared from sight. And at the end of that day also, Qujavarssuk came home again, as was his custom, with a catch of two seal.

Now by this time the Tupilak was fearfully hungry. But a Tupilak can only eat men, and therefore it now thought thus:

"Next time, I will go up on land and eat him there."

Then it swam over towards land, and as the shore was level, it moved swiftly, so as to come well up. But it struck its head on the ground, so that the pain pierced to its backbone, and when it tried to see what was there, the shore had changed to a steep cliff, and on the top of the cliff stood Qujavarssuk, all easy to see. Again it tried to swim up on to the land, but only hurt itself the more. And now it was surprised, and looked in vain for Qujavarssuk's house, for it could not see the house at all. And it was still lying there and staring up, when it saw that a great stone was about to fall on it, and hardly had it dived under water when the stone struck it, and broke a rib. Then it swam out and looked again towards land, and saw Qujavarssuk again quite clearly, and also his house.

Now the Tupilak thought:

"I must try another way. Perhaps it will be better to go through the earth."

And when it tried to go through the earth, so much was easy; it only remained then to come up through the floor of the house. But the floor of the house was hard, and not to be got through. Therefore it tried behind the house, and there it was quite soft. It came up there, and went to the passage way, and there was a big black bird, sitting there eating something. The Tupilak thought:

"That is a fortunate being, which can sit and eat."

Then it tried to get up over the walls at the back part of the house, by taking hold of the grass in the turf blocks. But when it got there, the bird's food was the only thing it saw. Again it tried to get a little farther, seeing that the bird appeared not to heed it at all, but then suddenly the bird turned and bit a hole just above its flipper. And this was very painful, so that the Tupilak floundered about with pain, and floundered about till it came right out into the water.

And because of all these happenings, it had now become so angered that it swam back at once to the man who had made it, in order to eat him up. And when it came there, he was sitting in his kayak with his face turned towards the sun, and telling no other thing than of the Tupilak which he had made. For a long time the Tupilak lay there beneath him, and looked at him, until there came this thought:

"Why did he make me a Tupilak, when afterwards

all the trouble was to come upon me?"

Then it swam up and attacked the kayak, and the water was coloured red with blood as it ate him. And having thus found food, the Tupilak felt well and strong and very cheerful, until at last it began to think thus:

"All the other Tupilaks will certainly call this a shameful thing, that I should have killed the one who made me."

And it was now so troubled with shame at this that it swam far out into the open sea and was never seen again. And men say that it was because of shame it did so.

One day the old one said to Qujavarssuk:

"You are named after a man who died of hunger at Amerdloq." It is told of the people of Amerdloq that they catch nothing but turbot.

And Qujavarssuk went to Amerdloq and lived

there with an old man, and while he lived there, he made always the same catch as was his custom. At last the people of Amerdloq began to say to one another:

"This must be the first time there have been so many black seal here in our country; every time he goes hunting he catches two seal."

At last one of the big hunters went out hunting with him. They fixed the heads to their harpoons, and when they had come a little way out from land, Qujavarssuk stopped. Then when the other had gone a little distance from him, he turned, and saw that Qujavarssuk had already struck one seal. Then he rowed towards him, but when he came up, it was already killed. So he left him again for a little while, and when he turned, Qujavarssuk had again struck. Then Qujavarssuk rowed home. And the other stayed out the whole day, but did not see a single seal.

When Qujavarssuk had thus continued as a great hunter, his mother said to him at last that he should marry. He gave her no answer, and therefore she began to look about herself for a girl for him to marry, but it was her wish that the girl might be a great glutton, so that there might not be too much lost of all that meat. And she began to ask all the unmarried women to come and visit her. And because of this there came one day a young woman who was not very beautiful. And this one she liked very much, for that she was a clever eater, and having regard to this, she chose her out as the one her son should marry. One day she said to her son:

"That woman is the one you must have." And her son obeyed her, as was his custom.

Every day after their marriage, the strongest man in Amerdloq called in at the window:

"Qujavarssuk! Let us see which of us can first get a bladder float for hunting the whale."

Qujavarssuk made no answer, as was his custom, but the old man said to him:

"We use only speckled skin for whales. And they are now at this time in the mouth of the river." After this, they went to rest.

Qujavarssuk slept, and awoke, and got up, and went away to the north. And when he had gone a little way to the north, he came to the mouth of a small fjord. He looked round and saw a speckled seal that had come up to breathe. When it went down again, he rowed up on the landward side of it, and fixed the head and line to his harpoon. When it came up again to breathe, he rowed to where it was, and harpooned it, and after this, he at once rowed home with it.

The old man made the skin ready, and hung it up behind the house. But while it was hanging there, there came very often a noise as from the bladder float, and this although there was no one there. This thing the old man did not like at all.

When the winter was coming near, the old man said one day to Qujavarssuk:

"Now that time will soon be here when the whales come in to the coast."

One night Qujavarssuk had gone out of the house, when he heard a sound of deep breathing from the west, and this came nearer. And because this was the first time he had heard so mighty a breathing, he went in and told the matter in a little voice to his wife. And he had hardly told her this, when the old man, whom he had thought asleep, said:

"What is that you are saying?"

"Mighty breathings which I have heard, and did not know them, and they do not move from that side where the sun is." This said Qujavarssuk.

The old one put on his boots, and went out, and came in again, and said:

"It is the breathing of a whale."

In the morning, before it was yet light, there came a sound of running, and then one came and called through the window:

"Qujavarssuk! I was the first who heard the whales breathing."

It was the strong man, who wished to surpass him in this. Qujavarssuk said nothing, as was his custom, but the old man said:

"Qujavarssuk heard that while it was yet night." And they heard him laugh and go away.

The strong man had already got out the umiak* into the water to row out to the whale. And then Qujavarssuk came out, and they had already rowed away when Qujavarssuk got his boat into the water. He got it full of water, and drew it up again on to the shore, and turned the stem in towards land and poured the water out, and for the second time he drew it down into the water. And not until now did he begin to look about for rowers. They went

**Umiak: a large boat, as distinct from the small kayak.*

out, and when they could see ahead, the strong man of Amerdloq was already far away. Before he had come up to where he was, Qujavarssuk told his rowers to stop and be still. But they wished to go yet farther, believing that the whale would never come up to breathe in that place. Therefore he said to them:

"You shall see it when it comes up."

Hardly had the umiak stopped still, when Qujavarssuk began to tremble all over. When he turned round, there was already a whale quite near, and now his rowers begged him eagerly to steer to where it was. But Qujavarssuk now saw such a beast for the first time in his life. And he said:

"Let us look at it."

And his rowers had to stay still. When the strong man of Amerdloq heard the breathing of the whale, he looked round after it, and there lay the beast like a great rock close beside Qujavarssuk. And he called out to him from the place where he was:

"Harpoon it!"

Qujavarssuk made no answer, but his rowers were now even more eager than before. When the whale had breathed long enough, it went down again. Now his rowers wished very much to go farther out, because it was not likely that it would come up again in that way the next time. But Qujavarssuk would not move at all.

The whale stayed a long time under the water, and when it came up again it was still nearer. Now Qujavarssuk looked at it again for a long time, and now his rowers became very angry with him at last. Not until it seemed that the whale must soon go down again did Qujavarssuk say:

"Now row towards it."

And they rowed towards it, and he harpooned it. And when it now floundered about in pain and went down, he threw out his bladder float, and it was not strange that this went under water at once.

And those farther out called to him now and said:

"When a whale is struck it will always swim out to sea. Row now to the place where it would seem that it must come up."

But Qujavarssuk did not answer, and did not move from the place where he was. Not until they called to him for the third time did he answer:

"The beasts I have struck move always farther in, towards my house."

And now they had just begun laughing at him out there, when they heard a washing of water closer in to shore, and there it lay, quite like a tiny fish, turning about in its death struggle. They rowed up to it at once and made a tow line fast. The strong man rowed up to them, and when he came to where they were, no one of them was eating. Then he said:

"Not one of you eating, and here a newly-killed whale?"

When he said this, Qujavarssuk answered:

"None may eat of it until my mother has first eaten."

But the strong man tried then to take a mouthful, although this had been said. And when he did so, froth came out of his mouth at once. And he spat out that mouthful, because it was destroying his mouth.

And they brought that catch home, and Qujavarssuk's mother ate of it, and then at last all ate of it likewise, and then none had any badness in the mouth from eating of it. But the strong man sat for a long time the only one of them all who did not eat, and that because he must wait till his mouth was well again.

And the strong man of Amerdloq did not catch a

whale at all until after Qujavarssuk had caught another one.

For a whole year Qujavarssuk stayed at Amerdloq, and when it was spring, he went back southward to his home. He came to his own land, and there at a later time he died.

And that is all.

KUNIGSEQ

THERE was once a wizard whose name was Kunigseq. One day, when he was about to call on his helping spirits and make a flight down into the underworld, he gave orders that the floor should be swilled with salt water, to take off the evil smell which might otherwise frighten his helping spirits away.

Then he began to call upon his helping spirits, and without moving his body, began to pass downward through the floor.

And down he went. On his way he came to a reef, which was covered with weed, and therefore so slippery that none could pass that way. And as he could not pass, his helping spirit lay down beside him, and by placing his foot upon the spirit, he was able to pass.

And on he went, and came to a great slope covered with heather. Far down in the underworld, men say, the land is level, and the hills are small; there is sun down there, and the sky is also like that which we see from the earth.

Suddenly he heard one crying: "Here comes Kunigseq."

By the side of a little river he saw some children looking for grey-fish.

And before he had reached the houses of men, he

met his mother, who had gone out to gather berries. When he came up to her, she tried again and again to kiss him, but his helping spirit thrust her aside.

"He is only here on a visit," said the spirit.

Then she offered him some berries, and these he was about to put in his mouth, when the spirit said:

"If you eat of them, you will never return."

A little after, he caught sight of his dead brother, and then his mother said:

"Why do you wish to return to earth again? Your kin are here. And look down on the sea-shore; see the great stores of dried meat. Many seal are caught here, and it is a good place to be; there is no snow, and a beautiful open sea."

The sea lay smooth, without the slightest wind. Two kayaks were rowing towards land. Now and again they threw their bird darts, and they could be heard to laugh.

"I will come again when I die," said Kunigseq.

Some kayaks lay drying on a little island; they were those of men who had just lost their lives when out in their kayaks.

And it is told that the people of the underworld said to Kunigseq:

"When you return to earth, send us some ice, for we thirst for cold water down here."

After that, Kunigseq went back to earth, but it is said that his son fell sick soon afterwards, and died. And then Kunigseq did not care to live any longer, having seen what it was like in the under world. So he rowed out in his kayak, and caught a guillemot, and a little after, he caught a raven, and having eaten these one after the other, he died. And then they threw him out into the sea.

THE WOMAN WHO HAD A BEAR AS A FOSTER-SON

THERE was once an old woman living in a place where others lived. She lived nearest the shore, and when those who lived in houses up above had been out hunting, they gave her both meat and blubber.

And once they were out hunting as usual, and now and again they got a bear, so that they frequently ate bear's meat. And they came home with a whole bear. The old woman received a piece from the ribs as her share, and took it home to her house. After she had come home to her house, the wife of the man who had killed the bear came to the window and said:

"Dear little old woman in there, would you like to have a bear's cub?"

And the old woman went and fetched it, and

brought it into her house, shifted her lamp, and placed the cub, because it was frozen, up on to the drying frame to thaw. Suddenly she noticed that it moved a little, and took it down to warm it. Then she roasted some blubber, for she had heard that bears lived on blubber, and in this way she fed it from that time onwards, giving it greaves to eat and melted blubber to drink, and it lay beside her at night.

And after it had begun to lie beside her at night it grew very fast, and she began to talk to it in human speech, and thus it gained the mind of a human being, and when it wished to ask its foster-mother for food, it would sniff.

The old woman now no longer suffered want, and those living near brought her food for the cub. The children came sometimes to play with it, but then the old woman would say:

"Little bear, remember to sheathe your claws when you play with them."

In the morning, the children would come to the window and call in:

"Little bear, come out and play with us, for now we are going to play."

And when they went out to play together, it would break the children's toy harpoons to pieces, but whenever it wanted to give any one of the children a push, it would always sheathe its claws. But at last it grew so strong, that it nearly always made the children cry. And when it had grown so strong the grown-up people began to play with it, and they helped the old woman in this way, in making the bear grow stronger. But after a time not even grown men dared play with it, so great was its strength, and then they said to one another:

"Let us take it with us when we go out hunting. It may help us to find seal."

And so one day in the dawn, they came to the old woman's window and cried:

"Little bear, come and earn a share of our catch; come out hunting with us, bear."

But before the bear went out, it sniffed at the old woman. And then it went out with the men.

On the way, one of the men said:

"Little bear, you must keep down wind, for if you do not so, the game will scent you, and take fright."

One day when they had been out hunting and were returning home, they called in to the old woman:

"It was very nearly killed by the hunters from the northward; we hardly managed to save it alive. Give therefore some mark by which it may be known; a broad collar of plaited sinews about its neck."

And so the old foster-mother made a mark for it to wear; a collar of plaited sinews, as broad as a harpoon line.

And after that it never failed to catch seal, and was stronger even than the strongest of hunters, and never stayed at home even in the worst of all weather. Also it was not bigger than an ordinary bear. All the people in the other villages knew it now, and although they sometimes came near to catching it, they would always let it go as soon as they saw its collar.

But now the people from beyond Angmagssalik heard that there was a bear which could not be caught, and then one of them said:

"If ever I see it, I will kill it."

But the others said:

"You must not do that; the bear's foster-mother could ill manage without its help. If you see it, do not harm it, but leave it alone, as soon as you see its mark."

One day when the bear came home as usual from hunting, the old foster-mother said:

"Whenever you meet with men, treat them as if you were of one kin with them; never seek to harm them unless they first attack."

And it heard the foster-mother's words and did as she had said.

And thus the old foster-mother kept the bear with her. In the summer it went out hunting in the sea, and in winter on the ice, and the other hunters now learned to know it's ways, and received shares of its catch.

Once during a storm the bear was away hunting as usual, and did not come home until evening. Then it sniffed at its foster-mother and sprang up on to the bench, where its place was on the southern side. Then the old foster-mother went out of the house, and found outside the body of a dead man, which the bear had hauled home. Then without going in again, the old woman went hurrying to the nearest house, and cried at the window:

"Are you all at home?"

"Why?"

"The little bear has come home with a dead man, one whom I do not know."

When it grew light, they went out and saw that it was the man from the north, and they could see he had been running fast, for he had drawn off his furs, and was in his underbreeches. Afterwards they heard that it was his comrades who had urged the bear to resistance, because he would not leave it alone.

A long time after this had happened, the old foster-mother said to the bear:

"You had better not stay with me here always; you will be killed if you do, and that would be a pity. You had better leave me."

And she wept as she said this. But the bear thrust its muzzle right down to the floor and wept, so greatly did it grieve to go away from her.

After this, the foster-mother went out every morning as soon as dawn appeared, to look at the weather, and if there were but a cloud as big as one's hand in the sky, she said nothing.

But one morning when she went out, there was not even a cloud as big as a hand, and so she came in and said:

"Little bear, now you had better go; you have your own kin far away out there."

But when the bear was ready to set out, the old foster-mother, weeping very much, dipped her hands in oil and smeared them with soot, and stroked the bear's side as it took leave of her, but in such manner that it could not see what she was doing. The bear sniffed at her and went away. But the old foster-mother wept all through that day, and her fellows in the place mourned also for the loss of their bear.

But men say that far to the north, when many bears are abroad, there will sometimes come a bear as big as an iceberg, with a black spot on its side.

Here ends this story.

IMARASUGSSUAQ, WHO ATE HIS WIVES

IT is said that the great Imarasugssuaq was wont to eat his wives. He fattened them up, giving them nothing but salmon to eat, and nothing at all to drink. Once when he had just lost his wife in the usual way, he took to wife the sister of many brothers, and her name was Misana. And after having taken her to wife, he began fattening her up as usual.

One day her husband was out in his kayak. And she had grown so fat that she could hardly move, but now she managed with difficulty to tumble down from the bench to the floor, crawled to the entrance, dropped down into the passage way, and began licking the snow which had drifted in. She licked and licked at it, and at last she began to feel herself lighter, and better able to move. And in this way she afterwards went out and licked up snow whenever her husband was out in his kayak, and at last she was once more quite able to move about.

One day when her husband was out in his kayak as usual, she took her breeches and tunic, and stuffed them out until the thing looked like a real human being, and then she said to them:

"When my husband comes and tells you to come out, answer him with these words: I cannot move because I am grown so fat. And when he then comes in and harpoons you, remember then to shriek as if in pain."

And after she had said these words, she began digging a hole at the back of the house, and when it was big enough, she crept in.

"Bring up the birds I have caught!"

But the dummy answered:

"I can no longer move, for I am grown so fat."

Now the dummy was sitting behind the lamp. And the husband coming in, harpooned that dummy wife with his great bird-spear. And the thing shrieked as if with pain and fell down. But when he looked closer, there was no blood to be seen, nothing but some stuffed-out clothes. And where was his wife?

And now he began to search for her, and as soon as he had gone out, she crept forth from her hiding-place, and took to flight. And while she was thus making her escape, her husband came after her, and seeing that he came nearer and nearer, at last she said:

"Now I remember, my amulet is a piece of wood."

And hardly had she said these words, when she was changed into a piece of wood, and her husband could not find her. He looked about as hard as ever he could, but could see nothing beyond a piece of wood anywhere. And he stabbed at that once or twice with his knife, but she felt no more than a little stinging pain. Then he went back home to fetch his axe, and then, as soon as he was out of sight, she changed back into a woman again and fled away to her brothers.

When she came to their house, she hid herself behind the skin hangings, and after she had placed herself there, her husband was heard approaching, weeping because he had lost his wife. He stayed there with them, and in the evening, the brothers began singing songs in mockery of him, and turning towards him also, they said:

"Men say that Imarasugssuaq eats his wives."

"Who has said that?"

"Misana has said that."

"I said it, and I ran away because you tried to kill me," said she from behind the hangings.

And then the many brothers fell upon Imarasugssuaq and held him fast that his wife might kill him; she took her knife, but each time she tried to strike, the knife only grazed his skin, for her fingers lost their power.

And she was still standing there trying in vain to stab him, when they saw that he was already dead. Here ends this story.

QALAGANGUASE, WHO PASSED TO THE LAND OF GHOSTS

THERE was once a boy whose name was Qalaganguase; his parents lived at a place where the tides were strong. And one day they ate seaweed, and died of it. Then there was only one sister to look after Qalaganguase, but it was not long before she also died, and then there were only strangers to look after him.

Qalaganguase was without strength, the lower part of his body was dead, and one day when the others had gone out hunting, he was left alone in the house. He was sitting there quite alone, when suddenly he heard a sound. Now he was afraid, and with great pains he managed to drag himself out of the house into the one beside it, and here he found a hiding-place behind the skin hangings. And while he was in hiding there, he heard a noise again, and in walked a ghost.

"Ai! There are people here!"

The ghost went over to the water tub and drank, emptying the dipper twice.

"Thanks for the drink which I thirsty one received," said the ghost. "Thus I was wont to drink when I lived on earth. And then it went out."

Now the boy heard his fellow-villagers coming up and gathering outside the house, and then they began to crawl in through the passage way.

"Qalaganguase is not here," they said when they came inside.

"Yes, he is," said the boy. "I hid in here because a ghost came in. It drank from the water tub there."

And when they went to look at the water tub, they saw that something had been drinking from it.

Then some time after, it happened again that the people were all out hunting, and Qalaganguase alone in the place. And there he sat in the house all alone, when suddenly the walls and frame of the house began to shake, and next moment a crowd of ghosts came tumbling into the house, one after the other, and the last was one whom he knew, for it was his sister, who had died but a little time before.

And now the ghosts sat about on the floor and began playing; they wrestled, and told stories, and laughed all the time.

At first Qalaganguase was afraid of them, but at last he found it a pleasant thing to make the night pass. And not until the villagers could be heard returning did they hasten away.

"Now mind you do not tell tales," said the ghost, "for if you do as we say, then you will gain strength again, and there will be nothing you cannot do." And one by one they tumbled out of the passage way. Only Qalaganguase's sister could hardly get out, and that was because her brother had been minding her little child, and his touch stayed her. And the hunters were coming back, and quite close, when she slipped out. One could just see the shadow of a pair of feet.

"What was that," said one. "It looked like a pair of feet vanishing away."

"Listen, and I will tell you," said Qalaganguase, who already felt his strength returning. "The house has been full of people, and they made the night pass pleasantly for me, and now, they say, I am to grow strong again."

But hardly had the boy said these words, when the strength slowly began to leave him.

"Qalaganguase is to be challenged to a singing contest," he heard them say as he lay there. And then they tied the boy to the frame post and let him swing backwards and forwards, as he tried to beat the drum. After that, they all made ready, and set out for their singing contest, and left the lame boy behind in the house all alone. And there he lay all alone, when his mother, who had died long since, came in with his father.

"Why are you here alone?" they asked.

"I am lame," said the boy, "and when the others went off to a singing contest, they left me behind."

"Come away with us," said his father and mother.

"It is better so, perhaps," said the boy.

And so they led him out, and bore him away to the land of ghosts, and so Qalaganguase became a ghost.

And it is said that Qalaganguase became a woman when they changed him to a ghost. But his fellow-villagers never saw him again.

ISIGALIGARSSIK

ISIGALIGARSSIK was a wifeless man, and he was very strong. One of the other men in his village was a wizard. Isigaligarssik was taken to live in a house with many brothers, and they were very fond of him.

When the wizard was about to call upon his spirits, it was his custom to call in through the window: "Only the married men may come and hear." And when they who were to hear the spirit calling went out, a little widow and her daughter and Isigaligarssik always stayed behind together in the house. Once, when all had gone out to hear the wizard, as was their custom, these three were thus left alone together. Isigaligarssik sat by the little lamp on the side bench, at work.

Suddenly he heard the widow's daughter saying something in her mother's ear, and then her mother turned towards him and said:

"This little girl would like to have you."

Isigaligarssik would also like to have her, and before the others of the house had come back, they were man and wife. Thus when the others of the house had finished and came back, Isigaligarssik had found a wife, and his house-fellows were very glad of this.

Next day, as soon as it was dark, one called, as was the custom:

"Let only those who have wives come and hear." And Isigaligarssik, who had before had no wife, felt now a great desire to go and hear this. But as soon as he had come in, the great wizard said to Isigaligarssik's wife:

"Come here; here."

When she had sat down, he told her to take off her shoes, and then he put them up on the drying frame. Then they made a spirit calling, and when that was ended, the wizard said to Isigaligarssik:

"Go away now; you will never have this dear little wife of yours again."

And then Isigaligarssik had to go home without a wife. And Isigaligarssik had to live without a wife. And every time there was a spirit calling, and he went in, the wizard would say:

"Ho, what are you doing here, you who have no wife?"

But now anger grew up slowly in him at this, and once when he came home, he said:

"That wizard in there has mocked me well, but next time he asks me, I shall know what to answer."

But the others of the village warned him, and said:

"No, no; you must not answer him. For if you answer him, then he will kill you."

But one evening when the bad wizard mocked him as usual Isigaligarssik said:

"Ho, and what of you who took my wife away?"

Now the wizard stood up at once, and when Isigaligarssik bent down towards the entrance to creep out, the wizard took a knife, and stabbed him with a great wound.

Isigaligarssik ran quickly home to his house, and said to his wife's mother:

"Go quickly now and take the dress I wore when I was little.* It is in the chest there."

And when she took it out, it was so small that it did not look like a dress at all, but it was very pretty. And he ordered her then to dip it in the water bucket. When it was wet, he was able to put it on, and when the lacing thong at the bottom touched the wound, it was healed.

Now when his house-fellows came out after the spirit-calling they thought to find him lying dead outside the entrance. They followed the blood-spoor, and at last he had gone into the house.

** The first dress worn by a child is supposed to act as a charm against wounds if the former wearer can put it on when a grown man.*

When they came in, he had not a single wound, and all were very glad for that he was healed again. And now he said:

"To-morrow I will go bow-shooting with him."

Then they slept, and awakened, and Isigaligarssik opened his little chest and searched it, and took out a bow that was so small it could hardly be seen in his hands. He strung that bow, and went out, and said:

"Come out now and see." Then they went out, and he went down to the wizard's house, and called through the window:

"Big man in there; come out now and let us shoot with the bow!" And when he had said this, he went and stood by a little river. When he turned to look round, the wizard was already by the passage of his house, aiming with his bow.

He said: "Come here." And then Isigaligarssik drew

up spittle in his mouth and spat straight down beside his feet.

"Come here," he said then to the great wizard. Then he went over to him, and came nearer and nearer, and stopped just before him. Now the wizard aimed with his bow towards him, and when he did this, the house-fellows cried to Isigaligarssik: "Make yourself small!" And he made himself so small that only his head could be seen moving backwards and forwards. The wizard shot and missed. And a second time he shot and missed.

Then Isigaligarssik stood up, and took the arrow, and broke it across and said:

"Go home; you cannot hit." And then the wizard went off, turning many times to look round. At last, when he bent down to get into his house through the passage way, Isigaligarssik aimed and shot at him. And they heard only the sound of his fall. The arrow was very little, and yet for all that it sent him all doubled up through the entrance, so that he fell down in the passage.

In this way Isigaligarssik won his wife again, and he lived with her afterwards until death.

THE INSECTS THAT WOOED A WIFELESS MAN

THERE was once a wifeless man. Yes, that is the way a story always begins. And it was his custom to run down to the girls whenever he saw them out playing. And the young girls always ran away from him into their houses.

And when the time of great hunting set in, and the kayak men lived in plenty, it always happened that he shamefully overslept himself every time he had made up his mind to go out hunting. He did not wake until the sun had gone down, and the hunters began to come in with their catch in tow.

One day when he awoke as usual about sunset, he got into his kayak all the same, and rowed off. Hardly had he passed out of sight of the houses, when he heard a man crying:

"My kayak has upset, help me."

And he rowed over and righted him again, and then he saw that it was one of the Noseless Ones, the people from beneath the earth.

"Now I will give you all my hide thongs with ornaments of walrus tusk," said the man who had upset.

"No," said the wifeless man; "such things I am not fit to receive; the only thing I cannot overcome is my miserable sleepiness."

"First come in with me to land," said the Fire Man. And they went in together.

When they reached the place, the Noseless One said:

"This is the man who saved my life when I was near to death."

"I happened to save you because my course lay athwart your own," said the wifeless man. "It is the first time for many days that I have been out at all in my kayak."

"One beast and one only you may choose when you are on your homeward way. And be careful never to tell what you have seen, or it will go ill with your hunting hereafter."

Those were the Fire Man's words. And then the wifeless man rowed home.

But when the time for his expected return had come, he was nowhere to be seen, and the young girls began to rejoice at the misfortune which must have befallen him. For they could not bear the sight of that man.

But then suddenly he came in sight round the point, and at once all cried:

"Here comes one who looks like the wifeless man."

And then all the young unmarried girls ran into their houses.

"And the wifeless man has made a catch," one cried.

And hardly had the evening begun to fall when the wifeless man went to rest, and hardly had the light appeared when the wifeless man went out hunting, long before his fellows. Hardly had the sun appeared in the sky, when the wifeless man came home with three seals. And his fellow-hunters were then but just preparing to set out.

Thus the days passed for that wifeless man. Early in the morning he would go out, and when the sun had only just begun to climb the sky, he would come home with his catch.

Then the unmarried girls began talking together.

"What has come to our wifeless man," they said, and began to vie with one another in seeking his favour.

"Let me, let me," they cried all together.

And the wifeless man turned towards them, and laughingly chose-out the best in the flock.

And now they lived together, the wifeless man and the girl, and every day there was freshly caught seal meat to be cut up. At last she grew weary, and cried:

"Why ever do you catch such a terrible lot?"

"Hmm," said he. "The seals come of themselves, and I catch them that is all."

But she kept on asking him, and so he said at last:

"It was in this way. Once . . ." But having said thus much, he ceased, and went to rest. But it was long before he could sleep. And the sun was just over the houses of the village before he awoke and set out next day.

That day he caught but one seal.

In the evening, his wife began again asking and asking, and seeing that she would not desist, at last he said:

"It was in this way. Once . . . well, I woke up in the evening, and rowed out, and heard a man crying for help, because his kayak had upset. And I rowed up to him and righted him again, and when I looked at him, it was one of the Noseless Ones."

"It was a good thing you were not idling about by the houses," said the Noseless One to me.

"I had but just got into my kayak," said I.

And thus he told all that had happened to him that day, and from that time forward he lost his power of hunting, for now his old sleepiness came over him once more, and he lost all.

At last he had not even skins enough to give his wife for her clothes, and so she ran away and left him. He set off in chase, but she escaped through a crevice in the rocks, a narrow place whereby he could just pass.

Now he lay in wait there, and soon he heard a whispering inside:

“You go out to him.”

And out crawled a blowfly, and said:

“Take me.”

“I will not take you,” said the wifeless man, “for you pick your food from the muck-heaps.”

The blowfly laughed and crawled back again, and he could hear it say:

“He will not take me, because I pick my food from the muck-heaps.”

Then there was more whispering inside.

"Now you go out."

And out came a fly.

"You may have me," it said.

"Thanks," said the wifeless man, "but I do not care for you at all. You lay your eggs about anyhow, and your eyes are quite abominably big."

At this the fly laughed, and went inside with the same message as before.

Again there was a whispering inside.

"Take me," said the cranefly.

"No, your legs are too long," said the wifeless man. And the cranefly went in again, laughing.

Then out came a centipede.

"Take me."

"I will not take you," said the wifeless man, "for you have far too many legs. Your body clings to the ground with all those legs, and your eyes are simply nasty."

And the centipede laughed a cackling laugh and went in again.

They whispered together again in there, and out came a gnat.

"Take me," said the gnat.

"No thanks, you bite," said the wifeless man. And the gnat went in again, laughing.

And then at last his wife bade him come in to her, since he would have none of the others, and at last he just managed to squeeze his body in through the crack, and then he took her to wife again.

"Comb my hair," said the wifeless man, now very

happy once more.

And his wife began, and said words above him thus:

"Do not wake until the fulmar begins to cry: sleep until we hear a sound of young birds."

And he fell asleep.

And when at last he awoke, he was all alone. The earth was blue with summer, and the fulmar cried noisily on the bird cliff. And it had been winter when he crawled in through the crack.

When he came down to his kayak, the skin was rotted through with age.

And then I suppose he reached home as usual, and now sits scratching himself at ease.

THE VERY OBSTINATE MAN

THERE was once an Obstinate Man no one in the world could be as obstinate as he. And no one dared come near him, so obstinate was he, and he would always have his own way in everything.

Once it came about that his wife was in mourning. Her little child had died, and therefore she was obliged to remain idle at home; this is the custom of the ignorant, and this we also had to do when we were as ignorant as they.

And while she sat thus idle and in mourning, her husband, that Obstinate One, came in one day and said:

"You must sew the skin of my kayak."

"You know that I am not permitted to touch any

kind of work," said his wife.

"You must sew the skin of my kayak," he said again. "Bring it down to the shore and sew it there."

And so the woman, for all her mourning, was forced to go down to the shore and sew the skin of her husband's kayak. But when she had been sewing a little, suddenly her thread began to make a little sound, and the little sound grew to a muttering, and louder and louder. And at last a monster came up out of the sea; a monster in the shape of a dog, and said:

"Why are you sewing, you who are still in mourning?"

"My husband will not listen to me, for he is so obstinate," she said.

And then the mighty dog sprang ashore and fell upon that husband.

But that Obstinate One was not abashed; as usual,

he thought he would get his own way, and his way now was to kill the dog. And they fought together, and the dog was killed.

But now the owner of the dog appeared, and he turned out to be the Moon Man.

And he fell upon that Obstinate One, but the Obstinate One would as usual not give way, but fell upon him in turn. He caught the Moon Man by the throat, and had nearly strangled him. He clenched and clenched, and the Moon Man was nearly strangled to death.

"There will be no more ebb-tide or flood if you strangle me," said the Moon Man.

But the Obstinate One cared little for that; he only clutched the tighter.

"The seal will never breed again if you strangle me," cried the Moon Man.

But the Obstinate One did not care at all, though

the Moon Man threatened more and more.

"There will never be dawn or daylight again if you kill me," said the Moon Man at last.

And at this the Obstinate One began to hesitate; he did not like the thought of living in the dark forever. And he let the Moon Man go.

Then the Moon Man called his dog to life again, and made ready to leave that place. And he took his team and cast the dogs up into the air one by one, and they never came down again, and at last there was the whole team of sledge dogs hovering in the air.

"May I come and visit you in the Moon?" asked the Obstinate One. For he suddenly felt a great desire to do so.

"Yes, come if you please," said the Moon Man. "But when you see a great rock in your way, take great care to drive round behind it. Do not pass it on the sunny side, for if you do, your heart will be torn out of you."

And then the Moon Man cracked his whip, and drove off through the naked air.

Now the Obstinate One began making ready for his journey to the moon. It had been his custom to keep his dogs inside the house, and therefore they had a thick layer of ingrown dirt in their coats. Now he took them and cast them out into the sea, that they might become clean again. The dogs, little used to going out at all, were nearly frozen to death by that cold water; they ran about, shivering with the cold.

Then the Obstinate One took a dog, and cast it up in the air, but it fell down heavily to earth again. He took another and did so, and then a third, but they all fell down again. They were still too dirty.

But the Obstinate One would not give in, and now he cast them out into the sea once more.

And when he then a second time tried casting

them up in the air, they stayed there. And now he made himself a sledge, threw his team up in the air, and drove off.

But when he came to the rock he was to drive round, this Obstinate One said to himself:

"Why should I drive round a rock at all? I will go by the sunny side."

When he came up alongside, he heard a woman singing drum songs, and whetting her knife; she kept on singing, and he could hear how the steel hummed as she worked.

Now he tried to overpower that old woman, but lost his senses. And when he came to himself, his heart was gone.

"I had better go round after all," he thought to himself. And he went round by the shady side.

Thus he came up to the moon, and told there how he had lost his heart merely for trying to drive round a rock by the sunny side.

Then the Moon Man bade him lie down at full length on his back, with a black sealskin under, which he spread on the floor.

This the Obstinate One did, and then the Moon Man fetched his heart from the woman and stuffed it in again.

And while he was there, the Moon Man took up one of the stones from the floor, and let him look down on to the earth. And there he saw his wife sitting on the bench, plaiting sinews for thread, and this although she was in mourning. A thick smoke rose from her body; the smoke of her evil thoughts. And her thoughts were evil because she was working before her mourning time was passed.

And her husband grew angry at this, forgetting that he had himself but newly bidden her work despite her mourning.

And after he had been there some time, the Moon

Man opened a stone in the entrance to the passage way, and let him look down. The place was full of walrus, there were so many that they had to lie one on top of another.

"It is a joy to catch such beasts," said the Moon Man, and the Obstinate One felt a great desire to harpoon one of them.

"But you must not, you cannot," said the Moon Man, and promised him a share of the catch he had just made himself. But the Obstinate One would not be content with this; he took harpoons from the Moon Man's store, and harpooned a walrus. Then he held it on the line, he was a man of very great strength, that Obstinate One and managed to kill it. And in the same way he also dealt with another.

After his return from the Moon Man's place, he left off being obstinate, and never again forced his wife to work while she was in mourning.

THE DWARFS

A MAN who was out in his kayak saw another kayak far off, and rowed up to it. When he came up with it, he saw that the man in it was a very little man, a dwarf.

"What do you want," asked the dwarf, who was very much afraid of the man.

"I saw you from afar and rowed up," said the man.

But the dwarf was plainly troubled and afraid.

"I was hunting a little fjord seal which I cannot hit," he said.

**The story-teller speaks the dwarf's part throughout in a hurried and jerky manner, to illustrate the little man's shyness.*

"Let me try," said the other. And so they waited until it came up to breathe. Hardly had it come up, when the harpoons went flying towards it, and entered in between its shoulder-blades.

"Ai, ai what a throw!" cried the dwarf in astonishment. And the man took the seal and made a tow-line fast.

Then the two kayaks set off together in towards land.

"Hum hum. Wouldn't care to . . . come and visit us?," said the dwarf suddenly.

But this the man would gladly do.

"Hum hum. I have a wife . . . and a daughter . . . very beautiful daughter . . . hum hum. Many men wanted her . . . wouldn't have them . . . can't take her by force . . . very strong. Thought of taking her to wife myself . . . hum hum. But she is too strong for me . . . own daughter."

They rowed on a while, and then the little one spoke again.

"Hum hum. Might perhaps do for you . . . you could manage her . . . what?"

"Let us first see her," said the man. And now they rowed into a great deep fjord.

When they came to the place, they landed and went up at once to the house of the little old man. And those in the house did all they could that the stranger might be well pleased. When they had been sitting there a while, the old man said:

"Hum hum . . . our guest has made a catch . . . he comes to us bringing game."

Now it was easy to see that they would gladly have tasted the flesh of that little seal. And so the guest said:

"If you care to cook that meat, then set to work

and cut it up as soon as you please. Cut it up and give to those who wish to eat of it."

The little old man was delighted at this, and sent out his two women-folk to cut up that seal. But they stayed away a long while, and no one came in with any meat. So the little old man went out to look for them.

And there stood the two women, hauling at the little fjord seal, which they could not manage to drag up from the shore. They could not even manage it with the old man s help. They hauled away, all three of them, bending their bodies to the ground in their efforts, but the seal would not move. Then at last the stranger came out, and he took that seal by the flipper with one hand, and carried it up that way.

"What strength, what strength!" The man is a giant indeed," cried the little folk. And they fell to work cutting up the seal, but to them it seemed as if they were cutting up a huge walrus, so hard did they find it to cut up that little seal.

And people came hurrying down from the houses up above, and all wished to share. The women of the house then shared out that seal. Each of the guests was given a little breastbone and no more, but this to them was a very great piece of meat. When they held such a piece in their hands, it reached to the ground, and their hands and clothes were covered with fat.

Inside on the bench sat an old hag who now began trying to make herself agreeable to the guest. She squeezed up close to him and kept on talking to him, and looking at him kindly. She was old and ugly, and the man would have nothing to do with her. Suddenly he gave a loud whistle.

"Ugh ugh!" cried the old hag in a fright, and fell down from the bench. Then she stumbled down into the passage way, and disappeared.

And now after they had feasted on the seal meat, those from the houses up above cried out:

"Let the guest now come up here; we have foxes liver to eat!"

And as he did not come at once, they cried again. And then he went up. The house was full of people, all busy eating foxes liver.

"It is very hard to cut," said the dwarfs. "It is dried."

And the dwarfs worked away as hard as they could, but could not cut it through. But the guest took and munched and crunched as if it had been fresh meat.

"Ai, ai see how he can eat," cried some.

But all those in the house were very kind to him, and would gladly have seen him married into their family. And the young women had dressed their hair daintily with mussel shells, that the guest might think them the finer. But he cared for none of them, for the little old man's daughter was the most beautiful.

And therefore he went down to that house again when it was time to go to rest. And he said he would have her to wife.

And so they lived happily together, and soon they had a child.

And now the man began to long for his own place and kin. He thought more and more of his old mother, who was still alive when he started off.

And so one day he said he was going to visit his home.

"We will all go with you," said the little old man; "we will visit your kinsfolk."

And so they made ready for the journey, and set out.

Now when they came to the place of real people, all these were greatly astonished to find their old comrade still alive. For they had thought him dead long since.

And the dwarf people lived happily enough among the real men, and after a little time they forgot to be troubled and afraid.

But one day when the little dwarf grandmother was sitting at the opening of the passage way with the little child, she dropped the child in the passage.

"Hlurp hlurp hlurp," was all she heard. A great dog, his face black on one side and white on the other, lay there in the passage, and it ate up the child on the spot.

"Ai ai," she cried. "Nothing is left but a little smear on the ground."

And now the dwarf folk were filled with horror, and the little old man was for setting off at once. So they gathered their belongings together and set out.

And whenever they came to a village, they went up on shore, and the old man always went up with his tent-skins on his back.

"Are there any dogs here? Is there a great beast with a black-and-white face?" was always the first thing he asked.

"Yes, indeed." And before they could turn round, the old man was back in his boat again, so great was his fear of dogs.

And at last the skin was worn quite away from his forehead with carrying of tent-skins up on to the shore in vain.*

One day they were lying-to, when a wind began to blow from the north.

"Are there dogs here?" asked the old man, and groaned, for his forehead was flayed and smarting,

**A heavy burden carried on the back is supported by a strap or thong passing over the forehead.*

so often had he borne those tent-skins up and down. But before any could answer, he heard the barking of the dogs themselves. And in a moment he was back in his boat again.

The wind had grown stronger. The seas were frothing white, and the foam was scattered about.

Then the old dwarf stood up in his boat and cried:

"The sky is clearing to the east with crested clouds."

Now this was a magic song, and as soon as he had sung it, the sea was calm and bright once more.

Then the old man went on again. So great was the power of his magic words that he could calm the sea. But for all that he had no peace, by reason of the dogs.

And he went on his way again, but whither he came at last I do not know.

THE BOY FROM THE BOTTOM OF THE SEA, WHO FRIGHTENED THE PEOPLE OF THE HOUSE TO DEATH

WELL, you see, it was the usual thing: "The Obstinate One" had taken a wife, and of course he beat her, and when he wanted to make it an extra special beating, he took a box, and banged her about with that.

One day, when he had been beating her as usual, she ran away. And she was just about to have a child at that time. She walked straight out into the sea, and was nearly drowned, but suddenly she came to herself again, and found that she was at the bottom of the sea. And there she built herself a house.

While she was down there, the child was born. And when she went to look at it, she nearly died of fright, it was so ugly. Its eyes were jellyfish, its hair of seaweed, and the mouth was like a mussel.

And now these two lived down there together. The child grew up, and when it was a little grown up, it could hear the children playing on the earth up above, and it said:

"I should like to go up and see."

"When you have grown stronger, then you may go," I said to his mother. And then the boy began practising feats of strength, with stones. And at last he was able to pick up stones as big as a chest, and carry them into the house.

One evening, when it was dark, they heard again a calling from above. The children, not content with simply shouting at their play, began crying out: "Iyoi-iyoi-iyoi" with all their might.

"Now I will go with you," said the mother. "But you must not go into the houses nearest the shore, for there I often fled in when your father would have beaten me; I have suffered much evil up there. And when you thrust in your head, be sure to look as angry as you can."

There were two houses on the shore, one a little way above the other. As they went up, the mother suddenly saw that her son was going into the one nearest the shore. And she cried:

"Ha-a; Ha-a! When your father beat me, I always ran in there. Go to the one up above."

And now the boy made his face fierce, and thrust in his head at the doorway, and all those inside fell down dead with fright. He would have beaten his father, but his father had died long since. Then he went down again to the bottom of the sea.

When the day dawned, the people from the house nearest the shore came out and said:

"Ai! What footsteps are these, all full of seaweed?" And seeing that the tracks led up to the house a little way above, they followed there, and found that all inside had died of fright.

THE RAVEN AND THE GOOSE

DO you know why the raven is so black, so dull and black in colour? It is all because of its own obstinacy. Now listen. It happened in the days when all the birds were getting their colours and the pattern in their coats. And the raven and the goose happened to meet, and they agreed to paint each other.

The raven began, and painted the other black, with a nice white pattern showing between.

The goose thought that very fine indeed, and began to do the same by the raven, painting it a coat exactly like its own.

But then the raven fell into a rage, and declared the pattern was frightfully ugly, and the goose, offended at all the fuss, simply splashed it black all over.

And now you know why the raven is black.

WHEN THE RAVENS COULD SPEAK

ONCE, long ago, there was a time when the ravens could talk. But the strange thing about the ravens speech was that their words had the opposite meaning. When they wanted to thank any one, they used words of abuse, and thus always said the reverse of what they meant.

But as they were thus so full of lies, there came one day an old man, and by magic means took away their power of speech. And since that time the ravens can do no more than shriek.

But the ravens nature has not changed, and to this day they are an ill-tempered, lying, thieving lot.

MAKITE

MAKITE, men say, took to wife the sister of many brothers, but he himself could never manage to catch a seal when he was out in his kayak. But his wife's brothers caught seal in great numbers. And so it was that one day he heard his wife say she would leave him, because he never caught anything. And in his grief at hearing this, he said to himself:

"This evening, when they are all asleep, I will go up into the hills and live there all alone."

When darkness had fallen, he set off up into the hills, but as he went, his wife's father, who was standing outside, saw him going, and cried in to the others in the house:

"Makite has gone up into the hills to live there all alone. Go after him."

The many brothers went out after him, but when they had nearly come up with him, he made his steps longer, and thus got farther and farther away from them, and at last they ceased to pursue him any more.

On his way he came to a house, and this was just as it was beginning to get light. He looked in, and saw that the hangings on the walls were of nothing but reindeer and foxes skins. And now he said to himself:

"Hmm, I may as well go in."

But as he went in, the hinge of the door creaked, and then another strange, deep sound was heard inside the house, and it began to shake.

At the same moment, the master of the house came in and said:

"Have you had nothing to eat yet?"

Makite said: "I will eat nothing until I know what are those things which look like candles, there in front of the window."

Then the lone-dweller said:

"That is no concern of one who is not himself a lone-dweller. Therefore he cannot tell you."

But then Makite said: "If you do not tell me, I will kill you."

And then at last he told.

"It may be you have seen to-day the great hills away in the blue to the south; if you go up to the top of the nearer hill, you will find nothing there, but he who climbs that one which lies farther away, and reaches the top, he will find such things there. But this cannot be done by one who is not a lone-dweller."

And not until he had said all this did Makite eat.

Then they both went to rest. And just as he was near falling asleep, the lone-dweller began to quiver slightly, but he pretended to sleep. And before Makite could see what he was about, the lone-dweller had strung his bow, and Makite, therefore, seeing he was preparing to kill him, pretended to wake up, and then the other laid aside his bow so quickly that it seemed as if he had not held anything at all. At last, when it was nearly dawn, the lone-dweller fell asleep, and then Makite tried very cautiously to get out, but as he was about to pass through the doorway, he again happened to draw the door to after him, and again it creaked as before with a strange sound. When he looked in through the window, the lone-dweller was about to get up.

Now Makite had laid his great spear a little way above the house, and he ran to the place. When he looked round, he saw that the man from the house was already in chase. Then he came to a big rock, and as there was no help for it, he commenced to run round. When he had run round it for the third time, he grasped his harpoon firmly, and without turning round, thrust it out behind him, and struck something soft. He had struck the other in the side.

Having now killed this one, and as there was no help for it, he wandered on at hazard, and came to a great plain. And in the middle of the plain was something which looked like a house. And he went up to it and found it was the house of a dwarf, and no end of people coming out of it. One went in and another came out, and so they kept on. He tried to get into the passage, but could not even get his foot in.

Then he heard someone inside saying:

"Heave up the passage way a little with your back, and then come in."

When he came in, it was a big place, and the old creature spoke to him, and said:

"When you go out, look towards the west; the inland-dwellers are coming."

And when Makite went out, he looked towards the west, and there he saw a great black thing approaching, and when he then came in again, the old man went to the window and called out:

"Here they are; they are close up now."

And then the dwarfs went out to fight, and took up their posts on the plain, one party opposite the other, and none said a word.

But suddenly the dog that was with the inland folk gave a great bark, and there came a mighty wave of water, rolling right up to the dwarfs.

But when it had come quite close to them, it suddenly grew quite small. And then the dwarfs' dog gave a bark. And at the same time the dwarfs' wave arose, and washed right up over the inland folk, and drowned them, and only few of them escaped alive.

When they came home again, Makite built himself

a house, and from the high hill fetched some of those things which looked like candles, and hung them up in his house. And he lived there in his house until he died.

And here ends this story.

ASALOQ

ASALOQ, men say, had a foster-brother. Once when he had come home after having been out in his kayak, his foster-brother had disappeared. He sought for him everywhere, but being unable to find him, he built a big umiak, and when it was built, he covered it with three layers of skins.

Then he rowed off southwards with his wife. And while they were rowing, they saw a black ripple on the sea ahead. When they came to the place, they saw that it was the sea-lice. And the outer most layer of skins on the boat was eaten away before they got through them.

Now they rowed onwards again, and saw once more a black ripple ahead. When they came to the place, they saw that it was the sea-serpents. And once again they slipped through with the loss of one layer of skins.

Having now but one layer of skins left, they went in great fear of what they might chance to meet next. But without seeing anything strange, they rounded a point, and came in sight of a place with many houses. Hardly had they come into land when the strangers caught hold of their boat, and hauled it up, so that Asaloq had no need to help.

And now it was learned that these were folk who had a strong man in their midst. Asaloq had been but a short time in one of the houses, when they heard the sound of one coming from outside and in through the passage way; it was the strong man's talebearer boy, and to make matters worse, a boy with a squint.

And now the people of the house said:

"Now that wretched boy will most certainly tell him you are here." And indeed, the boy was just about to run out again, when they caught hold of him and set him up behind the lamp. But hardly had they turned their backs on him for a moment, when he slipped out before any could move, and they heard the sound of his running footsteps in the snow without. And after a while, the window grew red with a constant filling of faces looking in to say: "We are sent to bid the stranger come."

And since there was no help for it, Asaloq went up there with them. When he came into the house, it was full of people, and he looked round and saw the strong man far in on the big bench. And at the moment Asaloq caught sight of him, the strong man said in a deep voice:

"Let us have a wrestling match."

And as he spoke, the others drew out a skin from

under the bench, and spread it on the floor. And after the skin had been spread out, food was brought in. And Asaloq ate till there was no more left. But as he rose, all that he had eaten fell out of his stomach. And then they began pulling arms.

And now Asaloq began mightily pulling the arms of all the men there, until the skin was worn from his arm, leaving the flesh almost bare.

And when he had straightened out all their arms, he went out of that house the strongest of all, and went out to his umiak and rowed away southwards with his wife. And when they had rowed a little way, they came to a little island, and pitched their tent on the sunny side. And when Asaloq then went up on the hillside to look out, he saw many umiaks coming from the northward, and they camped on the shady side. Then he heard them say:

"Now search carefully about." And others said:

"He can hardly be on such a little island."

And now Asaloq sang magic songs over them from the top of the hill, and at last he heard them say:

"We may as well go home again."

Now Asaloq stood and watched them row away, and not until they were out of sight did he set off again to the southward. At last they reached Aluk, and there their bones still rest.

Here ends this story.

UKALEQ

UKALEQ, men say, was a strong man. Whenever he heard news of game, even if it were a great bear, he had only to go out after it, and he never failed to kill it.

Once the winter came, and the ice grew firm, and then men began to go out hunting bears on the ice. One day there was a big bear. Ukaleq set off in chase, but he soon found that it was not to be easily brought down.

The bear sighted Ukaleq, and turned to pursue him. Ukaleq fled, but grew tired at length. Now and again he managed to wound the beast, but was killed himself at last, and at the same time the bear fell down dead.

Now when his comrades came to look at the bear, its teeth began to whisper, and then they knew that Ukaleq had been killed by a Magic Bear.* And as there was no help for it, they took the dead man home with them. And then his mother said:

"Lay him in the middle of the floor with a skin beneath him."

**A creature fashioned by an enemy, after the same manner as a Tupilak.*

She had kept the dress he had worn as a little child, and now that he was dead, she put it in her carrying bag, and went out with it to the cooking place in the passage. And when she got there, she said:

"For five days I will neither eat nor drink."

Then she began hushing the dress in the bag as if it were a child, and kept on hushing it until at last it began to move in the bag, and just as it had commenced to move, there came some out from the house and said:

"Ukaleq is beginning to quiver."

But she kept on hushing and hushing, and at last that which she had in the bag began trying to crawl out. But then there came one from the house and said:

"Ukaleq has begun to breathe; he is sitting up."

Hardly was this said when that which was in the bag sprang out, making the whole house shake. Then they made up a bed for Ukaleq on the side bench, and placed skins under him and made him sit up. And after five days had passed, and that without eating or drinking, he came to himself again, and commenced to go out hunting once more.

Then the winter came, and the winter was there, and the ice was over the sea, and when the ice had formed, they began to make spirit callings. The villages were close together, and all went visiting in other villages.

And at last Ukaleq set out with his family to a village near by, where there was to be a big spirit calling. The house where it was to be held was so big that there were three windows in it, and yet it was crowded with folk.

In the middle of the spirit calling, there was an old woman who was sitting cross-legged up on the bench, and she turned round towards the others and said:

"We heard last autumn that Ukaleq had been killed

by a Magic Bear." Hardly had she said those words when an old wifeless man turned towards her and said:

"Was it by any chance your Magic Bear that killed him?"

Then the old woman turned towards the others and said:

"Mine? Now where could I have kept such a thing?"

But after saying that she did not move. She even forgot to breathe, for shame at having been discovered by the wifeless man, and so she died on the spot.

After that Ukaleq went home, and never went out hunting bears again.

Here ends this story.

IKARDLITUARSSUK

IKARDLITUARSSUK, men say, had a little brother; they lived at a place where there were many other houses. One autumn the sea was frozen right out from the coast, without a speck of open water for a long way out. After this, there was great dearth and famine; at last their fellow-villagers began to offer a new kayak paddle as a reward for the one who should magic it away, but there was no wizard among the people of that village.

Then it came about that Ikardlituarssuk's little brother began to speak to him thus:

"Ikardlituarssuk, how very nice it would be to win that new paddle!"

And then it was revealed that Ikardlituarssuk had formerly sat on the knee of one of those present when the wizards called up their helping spirits.

Then it came about that Ikardlituarssuk one evening began to call upon his helping spirits. He called them up, and having called them up, went out, and having gone out, went down to the water's edge, crept in through a crack between the land and the ice, and started off, walking along the bottom of the sea.

He walked along, and when he came to seaweed, it seemed as if there lay dogs in among the weed. But these were sharks. Then on his way he saw a little house, and went towards it. When he came up to the entrance, it was narrow as the edge of a woman's knife. But he got in all the same, following that way which was narrow as the edge of a woman's knife. And when he came in, there sat the mother of Tornarssuk, the spirit who lived down there; she was sitting by her lamp and weeping. And picking behind her ears, she threw down many strange things. Inside her lamp were many birds that dived down, and inside the house were many seals that bobbed up.

And now he began tickling the weeping woman as hard as he could, to encourage her; and at last she was encouraged, and after this, she freed a number of the birds, and then made a sign to many of the seals to swim out of the house. And when they swam out, there was one of the fjord seals which she liked so much that she plucked a few of the hairs from its back, that she might have it to make breeches of when it was caught.

And when all this had been done, she went home, and went to rest without saying a word.

When they awoke next morning, the sea was quite dark ahead, and all the ice had gone. But when the villagers came out, she said to them:

"Do not kill more than one; if any of you should kill two, he will never kill again."

And furthermore she said:

"If any of you should catch a young fjord seal with a bare patch on its back, you must give it to me to make breeches."

When they came back, each of the hunters had made a catch; only one of them had caught two. And the man who had caught two seals that day never after caught any seal at all when he rowed out, but all the others always made a catch when they rowed out, and some of them even caught several at a time.

Thus it came about that Ikardlituarssuk with the little brother won the new paddle as a reward.

THE RAVEN WHO WANTED A WIFE

A LITTLE sparrow was mourning for her husband who was lost. She was very fond of him, for he caught worms for her.

As she sat there weeping, a raven came up to her

and asked:

"Why are you weeping?"

"I am weeping for my husband, who is lost; I was fond of him, because he caught worms for me," said the sparrow.

"It is not fitting for one to weep who can hop over high blades of grass," said the raven. "Take me for a husband; I have a fine high forehead, broad temples, a long beard and a big beak; you shall sleep under my wings, and I will give you lovely offal to eat."

"I will not take you for a husband, for you have a high forehead, broad temples, a long beard and a big beak, and will give me offal to eat."

So the raven flew away, flew off to seek a wife among the wild geese. And he was so lovesick that he could not sleep.

When he came to the wild geese, they were about

to fly away to other lands.

Said the raven to two of the geese:

"Seeing that a miserable sparrow has refused me, I will have you."

"We are just getting ready to fly away," said the geese.

"I will go too," said the raven.

"But consider this: that none can go with us who cannot swim or rest upon the surface of the water. For there are no icebergs along the way we go."

"It is nothing; I will sail through the air," said the raven.

And the wild geese flew away, and the raven with them. But very soon he felt himself sinking from weariness and lack of sleep.

"Something to rest on!" cried the raven, gasping.

"Sit you down side by side." And his two wives sat down together on the water, while their comrades flew on.

The raven sat down on them and fell asleep. But when his wives saw the other geese flying farther and farther away, they dropped that raven into the sea and flew off after them.

"Something to rest on!" gasped the raven, as it fell into the water. And at last it went to the bottom and was drowned.

And after a while, it broke up into little pieces, and its soul was turned into little "sea ravens." *

**A small black mollusk.*

THE MAN WHO TOOK A VIXEN TO WIFE

THERE was once a man who wished to have a wife unlike all other wives, and so he caught a little fox, a vixen, and took it home to his tent.

One day when he had been out hunting, he was surprised to find on his return that his little fox-wife had become a real woman. She had a lovely top-knot, made of that which had been her tail. And she had taken off the furry skin. And when he saw her thus, he thought her very beautiful indeed.

Now she began to talk about journeyings, and how greatly she desired to see other people. And so they went off, and came to a place and settled down there.

One of the men there had taken a little hare to wife. And now these two men thought it would be a pleasant thing to change wives. And so they did.

But the man who had borrowed the little vixen wife began to feel scorn of her after he had lived with her a little while. She had a foxy smell, and did not taste nice.

But when the little vixen noticed this she was very angry, for it was her great desire to be well thought of by the men. So she knocked out the lamp with her tail, dashed out of the house, and fled away far up into the hills.

Up in the hills she met a worm, and stayed with him.

But her husband, who was very fond of her, went out in search of her. And at last, after a long time, he found her living with the worm, who had taken human form.

But now it was revealed that this worm was the man's old enemy. For he had once, long before, burned a worm, and it was the soul of that worm which had now taken human form. He could even see the marks of burning in its face.

Now the worm challenged the man to pull arms, and they wrestled. But the man found the worm very easy to master, and soon he won.

After that he went out, no longer caring for his wife at all. And he wandered far, and came to the shore-dwellers. They had their houses on the shore, just by high-water mark.

Their houses were quite small, and the people themselves were dwarfs, who called the eider duck walrus. But they looked just like men, and were not in the least dangerous. We never see such folk nowadays, but our forefathers have told us about them, for they knew them.

And now when the man saw their house, which was roofed with stones, he went inside. But first he had to make himself quite small, though this of course was an easy matter for him, great wizard as he was.

As soon as he came in, they brought out meat to set before him. There was the whole fore-flipper of a mighty walrus. That is to say, it was really nothing more than the wing of an eider duck. And they fell to upon this and ate. But they did not eat it all up.

After he had stayed with these people some time he went back to his house. And I have no more to tell of him.

THE GREAT BEAR

A WOMAN ran away from her home because her child had died. On her way she came to a house. In the passage way there lay skins of bears. And she went in.

And now it was revealed that the people who lived in there were bears in human form.

Yet for all that she stayed with them. One big bear used to go out hunting to find food for them. It would put on its skin, and go out, and stay away for a long time, and always return with some catch or other. But one day the woman who had run away began to feel homesick, and greatly desired to see her kin. And then the bear spoke to her thus:

"Do not speak of us when you return to men," it said. For it was afraid lest its two cubs should be killed by the men.

Then the woman went home, and there she felt a great desire to tell what she had seen. And one day, as she sat with her husband in the house, she said to him:

"I have seen bears."

And now many sledges drove out, and when the bear saw them coming towards its house, it felt so sorry for its cubs that it bit them to death, that they might not fall into the hands of men.

But then it dashed out to find the woman who had betrayed it, and broke into her house and bit her to death. But when it came out, the dogs closed round it and fell upon it. The bear struck out at them, but suddenly all of them became wonderfully bright, and rose up to the sky in the form of stars. And it is these which we call Qilugtussat, the stars which look like barking dogs about a bear.

Since then, men have learned to beware of bears, for they hear what men say.

THE MAN WHO BECAME A STAR*

THERE was once an old man who stood out on the ice waiting for the seal to come up to their breathing holes to breathe. But on the shore, just

**The star is that which we know as Venus.*

opposite where he was, a crowd of children were playing in a ravine, and time after time they frightened away a seal just as he was about to harpoon it.

At last the old man grew angry with them for thus spoiling his catch, and cried out:

"Close up, Ravine, over those who are spoiling my hunting."

And at once the hillside closed over those children at play. One of them, who was carrying a little brother, had her fur coat torn.

Then they all fell to screaming inside the hill, for they could not come out. And none could bring them food, only water that they were able to pour down a crack, and this they licked up from the sides.

At last they all died of hunger.

And now the neighbours fell upon that old man

who had shut up the children by magic in the hill. He took to flight, and the others; ran after him.

But all at once he became bright, and rose up to heaven as a great star. We can see it now, in the west, when the lights begin to return after the great darkness. But it is low down, and never climbs high in the sky. And we call it Nalaussartoq, "he who stands and listens".

THE WOMAN WITH THE IRON TAIL

THERE was once a woman who had an iron tail. And more than this, she was also an eater of men. When a stranger came to visit her, she would wait until her guest had fallen asleep, and then she would jump up in the air, and fall down upon the sleeping one, who was thus pierced through by her tail.

Once there came a man to her house. And he lay down to sleep. And when she thought he had fallen asleep, she jumped up, and coming, over the place where he lay, dropped down upon him. But the man was not asleep at all, and he moved aside so that she fell down on a stone and broke her tail.

The man fled out to his kayak. And she ran after.

When she reached him, she cried:

"Oh, if I could only thrust my knife into him."

And as she cried, the man nearly upset for even her words had power.

"Oh, if only I could send my harpoon through her," cried the man in return. And so great was the power of his words that she fell down on the spot.

And then the man rowed away, and the woman never killed any one after that, for her tail was broken.

HOW THE FOG CAME

THERE was a Mountain Spirit, which stole corpses from their graves and ate them when it came home. And a man, wishing to see who did this thing, let himself be buried alive. The Spirit came, and saw the new grave, and dug up the body, and carried it off.

The man had stuck a flat stone in under his coat, in case the Spirit should try to stab him.

On the way, he caught hold of all the willow twigs whenever they passed any bushes, and made himself as heavy as he could, so that the Spirit was forced to put forth all its strength.

At last the Spirit reached its house, and flung down the body on the floor. And then, being weary, it lay down to sleep, while its wife went out to gather wood for the cooking.

"Father, father, he is opening his eyes," cried the children, when the dead man suddenly looked up.

"Nonsense, children, it is a dead body, which I have dropped many times among the twigs on the way," said the father.

But the man rose up, and killed the Mountain Spirit and its children, and fled away as fast as he could. The Mountain Spirit's wife saw him, and mistook him for her husband.

"Where are you going?" she cried.

The man did not answer, but fled on. And the woman, thinking something must be wrong, ran after him.

And as he was running over level ground, he cried:

"Rise up, hills!"

And at once many hills rose up.

Then the Mountain Spirit's wife lagged behind, having to climb up so many hills.

The man saw a little stream, and sprang across.

"Flow over your banks!" he cried to the stream. And now it was impossible for her to get across.

"How did you get across?" cried the woman.

"I drank up the water. Do you likewise." And the woman began gulping it down.

Then the man turned round towards her, and said:

"Look at the tail of your tunic; it is hanging down between your legs."

And when she bent down to look, her belly burst.

And as she burst, a steam rose up out of her, and

turned to fog which still floats about to this day among the hills.

THE MAN WHO AVENGED THE WIDOWS

THIS was in the old days, in those times when men were yet skilful rowers in kayaks. You know that there once came a great sickness which carried off all the older men, and the young men who were left alive did not know how to build kayaks, and thus it came about that the manner of hunting in kayaks was long forgotten.

But our forefathers were so skilful, that they would cross seas which we no longer dare to venture over. The weather also was in those times less violent than now; the winds came less suddenly, and it is said that the sea was never so rough.

In those times, there lived a man at Kangarssuk

whose name was Angusinanguaq, and he had a very beautiful wife, wherefore all men envied him. And one day, when they were setting out to hunt eider duck on the islands, the other men took counsel, and agreed to leave Angusinanguaq behind on a little lonely island there.

And so they sailed out to those islands, which lie far out at sea, and there they caught eider duck in snares, and gathered eggs, and were soon ready to turn homeward again. Then they pushed out from the land, without waiting for Angusinanguaq, who was up looking to his snares, and they took his kayak in tow, that he might never more be able to leave that island.

And now they hastened over towards the mainland. And the way was long.

But when they came in sight of the tents, they saw a man going from one tent to another, visiting the women whom they left behind at that place. They rowed faster, and came nearer. All the men of that place had gone out together for that hunting, and they could not guess who it might be that was now visiting among the tents.

Then an old man who was steering the boat shaded his eyes with his hand and looked over towards land.

"The man is Angusinanguaq," he said.

And now it was revealed that Angusinanguaq was a great wizard. When the umiaks had left, and he could not find his kayak, he had wound his body about with strips of hide, bending it into a curve, and then, as is the way of wizards, gathered magic power wherewith to move through the air. And thus he had come back to that place, long before those who had sought his death.

And from that day onwards, none ever planned again to take his wife. And it was well for them that they left him in peace.

For at that time, people were many, and there were people in all the lands round about. Out on the islands also there were people, and these were a fierce folk whom none might come near. Moreover when a kayak from the mainland came near their village, they would call down a fog upon him, so that he could not see, and in this manner cause him to perish.

But now one day Angusinanguaq planned to avenge his fellow-villagers. He rowed out to those unapproachable ones, and took them by surprise, being a great wizard, and killed many of the men, and cut off their heads and piled them up on the side bench. And having completed his revenge, he rowed away.

There was great joy among the widows of all those dead hunters when they learned that Angusinanguaq had avenged their husbands. And they went into his hut one by one and thanked him.

THE MAN WHO WENT OUT TO SEARCH FOR HIS SON

ONCE in the days of our forefathers, a man went out along the coasts, making search for his son. For that son had gone out in his kayak and had not returned.

One day he saw a giant beside a great glacier, and rowed up to him then. When he had entered the house, the giant drew forth a drum, a beautiful drum with a skin that had been taken from the belly of a man. Now the giant was about to give him this drum, but at the same time he felt such a violent desire to eat him up, that he trembled all over.

Just then some great salmon began dropping down through a hole in the roof, and the man was so frightened at this that he could scarcely eat. And he could not get out of the place.

But he was himself a great wizard, and now he began calling upon his helping spirits. And they were great.

"Killer whales, killer whales come forth, my helping spirits and show yourselves, for here is one who desires to eat me up."

And they came forth, and the house was crushed and the giant was killed, and the man set out again in search of his own.

Then he met another big man, and this man did nothing but eat men, and their kayaks he threw down into a great ravine. The man rowed up to this giant. And when he reached him, the man-eater said: "Come here and look," and led him to the deep ravine. And when the man looked down, the giant tried to thrust him backwards down into the depth.

But the man caught hold of the giant's legs and cast him down instead. And then he went on again.

And as he was rowing on, he heard the bone of a seal calling to him: "Take away the moss which has stopped up the hole that goes through me." And he did so, and went on again.

Another time he heard a mussel at the bottom of the sea crying:

"Here is a mussel that wishes to see you; come down to the bottom; row your kayak straight down through the water this way!"

That mussel wanted to eat him. But he did not heed it.

Then at last one day he saw an old woman, and rowed towards her, and came up to her. And she said:

"Let me dry your boots." And she took them and hung them up so high that he could not reach them. The man would have slept, but he could not sleep for fear.

"Give me my boots," he said. For it was now revealed that she was a man-eater. And so he got hold of his boots and fled down to his kayak, and the woman ran after him.

"If only I could catch him, and cut him up," she said. And as she spoke, the kayak nearly upset.

"If only I could send a bird dart through her," said the man. And as he spoke, the woman fell down on her back and broke her knife.

And then he rowed on his way. And on his way he met a man, and rowed up to him.

"See what a skin I have stretched out here," said the stranger. And he knew at once it was his son's kayak. The stranger had eaten his son, and there was his skin stretched out. The man therefore went up on land and trampled that man-eater to death, so that all his bones were crushed.

And then he went home again.

ATUNGAIT, WHO WENT A-WANDERING

ATUNGAIT, that great man, had once, it is said, a fancy to go out on a sledge trip with a strong woman. He took a ribbon seal and had it flayed, and forbade his wife to scrape the meat side clean, so that the skin might be as thick as possible. And so he had it dried.

When the winter had come, he went out to visit a tribe well known for their eagerness in playing football. He stayed among them for some time, and watched the games, carefully marking who was strongest among the players. And he saw that there was one among them a woman small of stature, who yet always contrived to snatch the ball from the others. Therefore he gave her the great thick skin he had brought with him, and told her to knead it soft. And this she did, though no other woman could have done it. Then he took her on his sledge and drove off on a wandering through the lands around.

On their way they came to a high and steep rock, rising up from the open water. Atungait sprang up on to that rock, and began running up it. So strong was he that at every step he bored his feet far down into the rock.

When he reached the top, he called to his dogs, and one by one they followed by the way of his footsteps, and reached the top, all of them save one, and that one died. And after that he hoisted up his sledge first, and then his wife after, and so they drove on their way.

After they had driven for some time, they came to a place of people. And the strange thing about these people was that they were all left-handed. And then they drove on again and came to some man-eaters; these ate one another, having no other food.

But they did not succeed in doing him any harm. And they drove on again and came to other people;

these had all one leg shorter than the other, and had been so from birth. They lay on the ground all day playing ajangat.* And they had a fine ajangat made of copper.

Atungait stayed there some time, and when the time came for him to set out once more, he stole their plaything and took it away with him, having first destroyed all their sledges.

But the lame ones, being unable to pursue, dealt magically with some rocky ridges, which then rushed over the ice towards the travellers.

Atungait heard something like the rushing of a river, and turning round, perceived those rocks rolling towards him.

"Have you a piece of sole-leather?" he asked his wife. And she had such a piece. She tied it to a string and let it drag behind the sledge. When the

**A game played with rings and a stick; the "ring and pin game."*

stones reached it, they stopped suddenly, and sank down through the ice. And the two drove on, hearing the cries of the lame ones behind them:

"Bring back our plaything, and give us our copper thing again."

But now Atungait began to long for his home, and not knowing in what part of the land they were, he told the woman with him to wait, while he himself flew off through the air. For he was a great wizard.

He soon found his house, and looked in through the window. And there sat his wife, rubbing noses with a strange man.

"Huh! You are not afraid of wearing away your nose, it seems." So he cried.

On hearing this, the wife rushed out of the house, and there she met her husband.

"You have grown clever at kissing," he said.

"No, I have not kissed any one," she cried.

Then Atungait grasped her roughly and killed her, because she had lied.

The strange man also came out now, and Atungait went towards him at once.

"You were kissing inside there, I see," he said.

"Yes," said the stranger. And Atungait let him live, because he spoke the truth.

And after that he flew back to the strong woman and made her his wife.

KUMAGDLAK AND THE LIVING ARROWS

KUMAGDLAK, men say, lived apart from his fellows. He had a wife, and she was the only living being in the place beside himself. One day his wife was out looking for stones to build a fireplace, and looking out over the sea, she saw many enemies approaching.

"An umiak and kayaks," she cried to her husband. And he was ill at ease on hearing this, for he lay in the house with a bad leg.

"My arrows bring my arrows!" he cried. And his wife saw that all his arrows lay there trembling. And that was because their points were made of the shinbones of men. And they trembled because their master was ill at ease.

Kumagdlak had made himself arrows, and feathered them with birds feathers. He was a great wizard, and by breathing with his own breath upon those arrows he could give them life, and cause them to fly towards his enemies and kill them. And when he himself stood unprotected before the weapons of his enemies, he would grasp the thong of the pouch in which his mother had carried him as a child, and strike out with it, and then all arrows aimed at him would fly wide of their mark. Now all the enemies hauled up on shore, and the eldest among them cried out:

"Kumagdlak! It is time for you to go out and taste the water in the land of the dead under the earth or perhaps you will go up into the sky?"

"That fate is more likely to be yours," answered Kumagdlak.

And standing at the entrance to his tent, he aimed at them with his bow. If but the first arrow could be sent whirling over the boats, then he knew that none of them would be able to harm him. He shot his arrow, and it flew over the boats. Then he aimed at the old man who had spoken, and that arrow cut through the string of the old man s bow, and pierced the old man himself. Then he began shooting down the others, his wife handing him the arrows as he shot. The men from the boats shot at him, but all their arrows flew wide. And his enemies grew fewer and fewer, and at last they fled.

And now Kumagdlak took all the bodies down by the shore and plundered them, taking their knives, and when the boats had got well out to sea, he called up a great storm, so that all the others perished.

But the waves washed the bodies this way and that along the coast, until the clothes were worn off them.

Here ends this story.

THE GIANT DOG

THERE was once a man who had a giant dog. It could swim in the sea, and was so big that it could haul whale and narwhal to shore. The narwhal it would hook on to its side teeth, and swim with them hanging there.

The man who owned it had cut holes in its jaws, and let in thongs through those holes, so that he could make it turn to either side by pulling at the thongs.

And when he and his wife desired to go journeying to any place, they had only to mount on its back.

The man had long wished to have a son, but as none was born to him, he gave his great dog the amulet which his son should have had. This amulet was a knot of hard wood, and the dog was thus made hard to resist the coming of death.

Once the dog ate a man, and then the owner of the dog was forced to leave that place and take land elsewhere. And while he was living in this new place, there came one day a kayak rowing in towards the land, and the man hastened to take up his dog, lest it should eat the stranger. He led it away far up into the hills, and gave it a great bone, that it might have something to gnaw at, and thus be kept busy.

But one day the dog smelt out the stranger, and came down from the hills, and then the man was forced to hide away the stranger and his kayak in a far place, lest the dog should tear them in pieces, for it was very fierce.

Now because the dog was so big and fierce, the man had many enemies. And once a stranger came driving in a sledge with three dogs as big as bears, to kill the giant dog. The man went out to meet that sledge, and the dog followed behind him. The dog pretended to be afraid at first, but then, when the stranger's dog set upon it in attack, it turned against them, and crushed the skulls of all three in its teeth.

After a time, the man noticed that his giant dog would go off, now and again, for long journeys in the hills, and would sometimes return with the leg of an inland-dweller. And now he understood that the dog had made it a custom to attack the inland-dwellers and bring back their legs to its master. He could see that the legs were legs of inland-dwellers, for they wore hairy boots.

And it is from this giant dog that the inland-dwellers got their great fear of all dogs. It would always appear suddenly at the window, and drag them out. But it was a good thing that something happened to frighten the inland-dwellers, for they had themselves an evil custom of carrying off lonely folk, especially women, when they had lost their way in the fog.

And that is all I know about the Giant Dog.

THE INLAND-DWELLERS OF ETAH

THERE came a sledge driving round to the east of Etah, up into the land, near the great lake. Suddenly the dogs scented something, and dashed off inland over a great plain. Then they checked, and sniffed at the ground. And now it was revealed that they were at the entrance to an inland-dweller's house.

The inland-dwellers screamed aloud with fear when they saw the dogs, and thrust out an old woman, but hurried in themselves to hide. The old woman died of fright when she saw the dogs.

Now the man went in, very ill at ease because he had caused the death of the old woman.

"It is a sad thing," he said, "that I should have caused you to lose that old one."

"It is nothing," answered the inland-dwellers; "her skin was already wrinkled; it does not matter at all."

Then the sledges drove home again, but the inland-dwellers were so terrified that they fled far up into the country.

Since then they have never been seen. The remains of their houses were all that could be found, and when men dug to see if anything else might be there, they found nothing but a single narwhal tusk.

The inland-dwellers are not really dangerous, they are only shy, and very greatly afraid of dogs. There was a woman of the coast-folk, Suagaq, who took a husband from among the inland folk, and when that husband came to visit her brothers, the blood sprang from his eyes at sight of their dogs.

And they train themselves to become swift runners, that they may catch foxes. When an inland-dweller is to become a swift runner, they stuff him into the skin of a ribbon seal, which is filled with worms, leaving only his head free. Then the worms suck all his blood, and this, they say, makes him very light on his feet.

There are still some inland-dwellers left, but they are now gone very far up inland.

THE MAN WHO STABBED HIS WIFE IN THE LEG

THERE was once a man whose name was Neruvkaq, and his wife was named Navarana, and she was of the tunerssuit, the inland-dwellers. She had many brothers, and was herself their only sister. And they lived at Natsivilik, the place where there is a great stone on which men lay out meat.

But Neruvkaq was cruel to his wife; he would stab her in the leg with an awl, and when the point reached her shinbone, she would snivel with pain.

"Do not touch me; I have many brothers," she said to her husband.

And as he did not cease from ill-treating her, she ran away to those brothers at last. And they were of the tunerssuit, the inland-dwellers.

Now all these many brothers moved down to Natsivilik, and when they reached the place, they sprang upon the roof of Neruvkaq's house and began to trample on it. One of them thrust his foot through the roof, and Neruvkaq's brother cut it off at the joint.

"He has cut off my leg," they heard him say. And then he hopped about on one leg until all the blood was gone from him and he died.

But Neruvkaq hastened to put on his tunic, and this was a tunic he had worn as a little child, and it had been made larger from time to time. Also it was covered with pieces of walrus tusk, sewn all about.

None could kill him as long as he wore that.

And now he wanted to get out of the house. He put the sealskin coat on his dog, and thrust it out. Those outside thought it was Neruvkaq himself, and stabbed the dog to death.

Neruvkaq came close on the heels of the dog, and jumped up to the great stone that is used to set out meat on. So strongly did he jump that his footmarks are seen on the stone to this day. Then he took his arrows all barbed with walrus tusk, and began shooting his enemies down.

His mother gave him strength by magic means.

Soon there were but few of his enemies left, and these fled away. They fled away to the southward, and fled and fled without stopping until they had gone a great way.

But Navarana, who was now afraid of her husband, crept in under the bench and hid herself there. And as she would not come out again, her husband thrust in a great piece of walrus meat, and she chewed and gnawed at it to her heart's content.

"Come out, come out, for I will never hurt you any more," he said. But she had grown so afraid of him that she never came out any more, and so she died where she was at last the old sneak!

THE SOUL THAT LIVED IN THE BODIES OF ALL BEASTS

THERE was a man whose name was Avovang. And of him it is said that nothing could wound him. And he lived at Kangerdlugssuaq.

At that time of the year when it is good to be out, and the days do not close with dark night, and all is nearing the great summer, Avovang's brother stood one day on the ice near the breathing hole of a seal.

And as he stood there, a sledge came dashing up, and as it reached him, the man who was in it said:

"There will come many sledges to kill your brother."

The brother now ran into the house to tell what he had heard. And then he ran up a steep rocky slope and hid away.

The sledges drove up before the house, and Avovang went out to meet them, but he took with him the skin of a dog's neck, which had been used to wrap him in when he was a child. And when then the men fell upon him, he simply placed that piece of skin on the ground and stood on it, and all his enemies could not wound him with their weapons, though they stabbed again and again.

At last he spoke, and said mockingly:

"All my body is now like a piece of knotty wood, with the scars of the wounds you gave me, and yet you could not bring about my death."

And as they could not wound him with their stabbing, they dragged him up to the top of a high cliff, thinking to cast him down. But each time they caught hold of him to cast him down, he changed himself into another man who was not their enemy. And at last they were forced to drive away, without having done what they wished.

It is also told of Avovang, that he once desired to travel to the south, and to the people who lived in the south, to buy wood. This men were wont to do in the old days, but now it is no longer so.

And so they set off, many sledges together, going southward to buy wood. And having done what they wished, they set out for home. On the way, they had made a halt to look for the breathing holes of seal, and while the men had been thus employed, the women had gone on. Avovang had taken a wife on that journey, from among the people of the south.

And while the men stood there looking for seal holes, all of them felt a great desire to possess Avovang's wife, and therefore they tried to kill him. Qautaq stabbed him in the eyes, and the others caught hold of him and sent him sliding down through a breathing hole into the sea.

When his wife saw this, she was angry, and taking the wood which they had brought from the south, she broke it all into small pieces. So angry was she at thus being made a widow.

Then she went home, after having spoiled the men's wood. But the sledges drove on.

Suddenly a great seal came up ahead of them, right in their way, where the ice was thin and slippery. And the sledges drove straight at it, but many fell through and were drowned at that hunting. And a little after, they again saw something in their way. It was a fox, and they set off in chase, but driving at furious speed up a mountain of screw-ice, they were dashed down and killed. Only two men escaped, and they made their way onward and told what had come to the rest.

And it was the soul of Avovang, whom nothing could wound, that had changed, first into a seal and then into a fox, and thus brought about the death of his enemies. And afterwards he made up his mind to let himself be born in the shape of every beast on earth, that he might one day tell his fellow-men the manner of their life.

At one time he was a dog and lived on meat which he stole from the houses. When he was pressed for food, he would carefully watch the men about the houses, and eat anything they threw away.

But Avovang soon tired of being a dog, on account of the many beatings which fell to his lot in that life. And so he made up his mind to become a reindeer.

At first he found it far from easy, for he could not keep pace with the other reindeer when they ran.

"How do you stretch your hind legs at a gallop?",

he asked one day.

"Kick out towards the farthest edge of the sky," they answered. And he did so, and then he was able to keep pace with them.

But at first he did not know what he should eat, and therefore he asked the others.

"Eat moss and lichen," they said.

And he soon grew fat, with thick suet on his back.

But one day the herd was attacked by a wolf, and all the reindeer dashed out into the sea, and there they met some kayaks in their flight, and one of the men killed Avovang.

He cut him up, and laid the meat in a cairn of stones. And there he lay, and when the winter came, he longed for the men to come and bring him home. And glad was he one day to hear the stones rattling down, and when they commenced to eat him, and cracked the bones with pieces of rock to get at the marrow, Avovang escaped and changed himself into a wolf.

And now he lived as a wolf, but here as before he found that he could not keep up with his comrades at a run. And they ate all the food, so that he got none.

"Kick up towards the sky," they told him. And then at once he was able to overtake all the reindeer, and thus get food.

And later he became a walrus, but found himself unable to dive down to the bottom; all he could do was to swim straight ahead through the water.

"Take off as if from the middle of the sky; that is what we do when we dive to the bottom," said the others. And so he swung his hindquarters up to the sky, and down he went to the bottom. And his comrades taught him what to eat; mussels and little white stones.

Once also he was a raven. "The ravens never lack

food," he said, "but they often feel cold about the feet."

Thus he lived the life of every beast on earth. And at last he became a seal again. And there he would lie under the ice, watching the men who came to catch him. And being a great wizard, he was able to hide himself away under the nail of a man's big toe.

But one day there came a man out hunting who had cut off the nail of his big toe. And that man harpooned him. Then they hauled him up on the ice and took him home.

Inside the house, they began cutting him up, and when the man cast the mittens to his wife, Avovang went with them, and crept into the body of the woman. And after a time he was born again, and became once more a man.

PAPIK, WHO KILLED HIS WIFE'S BROTHER

THERE was once a man whose name was Papik, and it was his custom to go out hunting with his wife's brother, whose name was Ailaq. But whenever those two went out hunting together, it was always Ailaq who came home with seal in tow, while Papik returned empty-handed. And day by day his envy grew.

Then one day it happened that Ailaq did not return at all. And Papik was silent at his home-coming.

At last, late in the evening, that old woman who was Ailaq's mother began to speak.

"You have killed Ailaq."

"No, I did not kill him," answered Papik.

Then the old woman rose up and cried:

"You killed him, and said no word. The day shall yet come when I will eat you alive, for you killed Ailaq, you and no other."

And now the old woman made ready to die, for it was as a ghost she thought to avenge her son. She took her bearskin coverlet over her, and went and sat down on the shore, close to the water, and let the tide come up and cover her.

For a long time after this, Papik did not go out hunting at all, so greatly did he fear the old woman's threat. But at last he ceased to think of the matter, and began to go out hunting as before.

One day two men stood out on the ice by the breathing holes. Papik had chosen his place a little farther off, and stood there alone. And then it came. They heard the snow creaking, with the sound of a cry, and the sound moved towards Papik, and a fog came down over the ice. And soon they heard shouts as of one in a fury, and the screaming of one in fear; the monster had fallen upon Papik, to devour him.

And now they fled in towards land, swerving wide to keep away from what was happening there. On their way, they met sledges with hunters setting out; they threw down their gear, and urged the others to return to their own place at once, lest they also should be slain by fear.

When they reached their village, all gathered together in one house. But soon they heard the monster coming nearer over the ice, and then all hurried to the entrance, and crowding together, grew yet more greatly stricken with fear. And pressing thus against each other, they struggled so hard that one fatherless boy was thrust aside and fell into a tub full of blood. When he got up, the blood poured from his clothes, and wherever they went, the snow was marked with blood.

"Now we are already made food for that monster," they cried, "since that wretched boy marks out the way with a trail of blood."

"Let us kill him, then," said one. But the others took pity on him, and let him live.

And now the evil spirit came in sight out on the ice; they could see the tips of its ears over the hummocks as it crept along. When it came up to the houses, not a dog barked, and none dared try to surround it, for it was not a real bear. But at last an old woman began crying to the dogs:

"See, there is your cousin, bark at him!" And now the dogs were loosed from the magic that bound them, and when the men saw this, they too dashed forward, and harpooned that thing.

But when they came to cut up the bear, they knew its skin for the old woman's coverlet, and its bones were human bones.

And now the sledges drove out to find the gear they had left behind, and they saw that everything was torn to pieces. And when they found Papik, he was cut about in every part. Eyes, nose and mouth and ears were hacked away, and the scalp torn from his head. Thus that old woman took vengeance for the killing of her son Ailaq.

And so it was our fathers used to tell: when any man killed his fellow without good cause, a monster would come and strike him dead with fear, and leave no part whole in all his body. The people of old times thought it an ill thing for men to kill each other.

This story I heard from the men who came to us from the far side of the great sea.

PATUSSORSSUAQ, WHO KILLED HIS UNCLE

THERE lived a woman at Kugkat, and she was very beautiful, and Alataq was he who had her to wife. And at the same place lived Patussorssuaq, and Alataq was his uncle. He also had a wife, but was yet fonder of his uncle's wife than of his own.

But one day in the spring, Alataq was going out on

a long hunting journey, and made up his mind to take his wife with him. They were standing at the edge of the ice, ready to start, when Patussorssuaq came down to them.

"Are you going away?" he asked. "Yes, both of us," answered Alataq.

But when Patussorssuaq heard thus, he fell upon his uncle and killed him at once, for he could not bear to see the woman go away. When Patussorssuaq's wife saw this, she snatched up her needle and sewing ring, and fled away, following the shadow of the tent, over the hills to the place where her parents lived. She had not even time to put on her skin stockings, and therefore her feet grew sore with treading the hills. On her way up inland she saw people running about with their hoods loose on their heads, as is the manner of the inland folk, but she had no dealings with them, for they fled away. Then, coming near at last to her own place, she saw an old man, and running up, she found it was her father, who was out in search of birds. And the two went gladly back to his tent.

Now when Patussorssuaq had killed his uncle, he at once went up to his own tent, thinking to kill his own wife, for he was already weary of her. But she had fled away.

Inside the tent sat a boy, and Patussorssuaq fell upon him, crying:

"Where is she? Where is she gone?"

"I have seen nothing, for I was asleep," cried the boy, speaking falsely because of his great fear. And so Patussorssuaq was forced to desist from seeking out his wife.

And now he went down and took Alataq's wife and lived with her. But after a little time, she died. And thus he had but little joy of the woman he had won by misdeed. And he himself was soon to suffer in another way.

At the beginning of the summer, many people were gathered at Natsivilik, and among them was Patussorssuaq. One day a strange thing happened to him, while he was out hunting: a fox snapped at the fringe of his coat, and he, thinking it to be but a common fox, struck out at it, but did not hit. And afterwards it was revealed that this was the soul of dead Alataq, playing with him a little before killing him outright. For Alataq's amulet was a fox.

And a little time after, he was bitten to death by the ghost of Alataq, coming upon him in the shape of a bear. His daughter, who was outside at that time, heard the cries, and went in to tell of what she had heard, but just as she came into the house, behold, she had quite forgotten all that she wished to say. And this was because that vengeful spirit had by magic means called down forgetfulness upon her.

Afterwards she remembered it, but then it was too late. They found Patussorssuaq torn to pieces, torn limb from limb; he had tried to defend himself with great pieces of ice, as they could see, but all in vain.

Thus punishment falls upon the man who kills.

THE MEN WHO CHANGED WIVES

THERE were once two men, Talilarssuaq and Navssarssuaq, and they changed wives. Talilarssuaq was a mischievous fellow, who was given to frightening people.

One evening, sitting in the house with the other's wife, whom he had borrowed, he thrust his knife suddenly through the skins of the bench. Then the woman ran away to her husband and said:

"Go in and kill Talilarssuaq; he is playing very dangerous tricks."

Then Navssarssuaq rose up without a word, and put on his best clothes, and took his knife, and went out. He went straight up to Talilarssuaq, who was now lying on the bench talking to himself, and pulled him out on the floor and stabbed him.

"You might at least have waited till I had dressed," said Talilarssuaq. But Navssarssuaq hauled him out through the passage way, cast him on the rubbish heap and went his way, saying nothing.

On the way he met his wife.

"Are you not going to murder me, too?" she asked.

"No," he answered in a deep voice. "For Pualuna is not yet grown big enough to be without you." Pualuna was their youngest son.

But some time after that deed he began to perceive that he was haunted by a spirit.

"There is some invisible thing which now and again catches hold of me," he said to his comrades. And that was the avenging spirit, watching him.

But about this time, many in the place fell sick.

And among them was Navssarssuaq. The sickness killed him, and thus the avenging spirit was not able to tear him in pieces.

ARTUK, WHO DID ALL FORBIDDEN THINGS

A MAN whose name was Artuk had buried his wife, but refused to remain aloof from doings which those who have been busied with the dead are forbidden to share. He said he did not hold by such old customs.

Some of his fellow-villagers were at work cutting up frozen meat for food. After watching them for a while as they worked at the meat with their knives, he took a stone axe and hacked at the meat, saying:

"That is the way to cut up meat."

And this he did although it was forbidden.

And on the same day he went out on to the ice and took off his inner coat to shake it, and this he did although it was forbidden.

Also he went up on to an iceberg and drank water which the sun had melted there, knowing well that this was likewise forbidden.

And all these things he did in scorn of that which his fellows believed. For he said it was all lies.

But one day when he was starting out with his sledge, fear came upon him, and he dared not go alone. And as his son would not go with him willingly, he took him, and bound him to the uprights of the sledge, and carried him so.

He never returned alive.

Late in the evening, his daughter heard in the air the mocking laughter of two spirits. And she knew at once that they were laughing so that she might know how her father had been punished for his ill-doing.

On the following day, many sledges went out to search for Artuk. And they found him, far out on the ice, torn to pieces, as is the way with those whom the spirits have punished for refusing to observe the customs of their forefathers. And the son, who was bound to the sledge, had not been touched, but he had died of fright.

THE THUNDER SPIRITS

TWO sisters, men say, were playing together, and their father could not bear to hear the noise they made, for he had but few children, and was thus not wont to hear any kind of noise. At last he began to scold them, and told them to go farther away with their playing.

When the girls grew up, and began to understand things, they desired to run away on account of their father's scolding. And at last they set out, taking with them only a little dogskin, and a piece of boot skin, and a fire stone. They went up into a high mountain to build themselves a house there.

Their father and mother made search for them in vain, for the girls kept hiding themselves; they had grown to be true mountain dwellers, keeping far from the places of men. Only the reindeer hunters saw them now and again, but the girls always refused to go back to their kin.

And when at last the time came when they must die of hunger, they turned into evil spirits, and became thunder.

When they shake their dried boot skin, then the gales come up, the south-westerly gales. And great fire is seen in the heavens whenever they strike their fire stone, and the rain pours down whenever they shed tears.

Their father held many spirit callings, hoping to make them return. But this he ceased to do when he found that they were dead.

But men say that after those girls had become spirits, they returned to the places of men, frightening many to death. They came first of all to their father and mother, because of the trouble they had made. The only one they did not kill was a woman bearing a child on her back. And they let her live, that she might tell how terrible they were. And tales are now told of how terrible they were.

When the thunder spirits come, even the earth itself is stricken with terror. And stones, even those which lie on level ground, and not on any slope at all, roll in fear towards men.

Thus the thunder comes with the south-westerly gales; there is a noise and crackling in the air, as of dry skins shaken, and the sky glows from time to time with the fire from their firestone. Great rocks, and everything which stands up high in the air, begin to glow.

When this happens, men use to take out a red dog, and cut its ear until the blood comes, and then lead the beast round about the house, letting the blood drip everywhere, for then the house will not take fire.

A red dog was the only thing they feared, those girls who were turned to thunder.

NERRIVIK

A BIRD once wished to marry a woman. He got himself a fine sealskin coat, and having weak eyes, made spectacles out of a walrus tusk, for he was greatly set upon looking as nice as possible. Then he set off, in the shape of a man, and coming to a village, took a wife, and brought her home.

Now he began to go out catching fish, which he called seal, and brought home to his wife.

Once it happened that he lost his spectacles, and his wife, seeing his bad eyes, burst out weeping, because he was so ugly.

But her husband only laughed. “Oho, so you saw my eyes? Hahaha!” And he put on his spectacles again.

Then her brothers, who longed for their sister, came out one day to visit her. And her husband being out hunting, they took her away with them. The husband was greatly distressed when he came home and found her gone, and thinking someone must have carried her off, he set out in pursuit. He swung his wings with mighty force, and raised a violent storm, for he was a great wizard.

When the storm came up, the boat began to take in water, and the wind grew fiercer, as he doubled the beating of his wings. The waves rose white with foam, and the boat was near turning-over. And when those in the boat began to suspect that the woman was the cause of the storm, they took her up and cast her into the sea. She tried to grasp the side of the boat, but then her grandfather sprang up and cut off her hand.

And so she was drowned. But at the bottom of the sea, she became Nerrivik, the ruler over all the creatures in the sea. And when men catch no seal, then the wizards go down to Nerrivik. Having but one hand, she cannot comb her hair, and this they do for her, and she, by way of thanks, sends seal and other creatures forth to men.

That is the story of the ruler of the sea. And men call her Nerrivik * because she gives them food.

* *Literally, "Meat Dish."*

THE WIFE WHO LIED

NAVARANAPALUK, men say, came of a tribe of man-eaters, but when she grew up, she was taken to wife by one of a tribe that did not eat men.

Once when she was going off on a visit to her own people, she put mittens on her feet instead of boots. And this she did in order to make it appear that her husband's people had dealt ill by her.

It was midwinter, and her kinsfolk pitied her greatly when they saw her come to them thus. And they agreed to make war against the tribe to which her husband belonged.

So they set out, and came to that village at a time when all the men were away, and only the women at home; these they took and slew, and only three escaped. One of them had covered herself with the skin which she was dressing when they came, the second had hidden herself in a box used for dog's meat, and the third had crept into a store shed.

When the men came home, they found all their womenfolk killed, and at once they thought of Navaranapaluk, who had fled away, And they were the more angered, that the slayers had hoisted the bodies of the women on long poles, with the points stuck through them.

They fell to at once making ready for war against those enemies, and prepared arrows in great numbers. The three women who were left alive plaited sinew thread to fix the points of the arrows; and so eagerly did they work that at last no more flesh was left on their fingers, and the naked bone showed through.

When all things were ready, they set out, and coming up behind the houses of their enemies, they hid themselves among great rocks.

The slayers had kept watch since their return, believing that the avengers would not fail to come, and the women took turns at the watching.

And now it is said that one old woman among them had a strange dream. She dreamed that two creatures were fighting above her head. And when she told the others of this, they all agreed that the avengers must be near. They gathered together in one house to ask counsel of the spirits, and when the spirit calling had commenced, then suddenly a dog upon the roof of the house began to bark.

The men dashed out, but their enemies had already surrounded the house, and now set about to take their full revenge, shooting down every man with arrows. At last, when they were no more left, they chose themselves wives from among the widows, and bore them off to their own place.

But two of them took Navaranapaluk and hurried off with her.

And she, thinking that both wished to have her to wife, cried out:

"Which is it to be? Which is it to be?"

The men laughed, and made no answer, but ran on with her.

Then suddenly they cut through both her arms with their knives. And soon she fell, and the blood went from her, and she died.

This fate they meted out to her because she lied.

KAGSSAGSSUK, THE HOMELESS BOY WHO BECAME A STRONG MAN

ONE day, it is said, when the men and women in the place had gone to a spirit calling, the children were left behind, all in one big house, where they played, making a great noise. A homeless boy named Kagssagssuk was walking about alone outside, and it is said that he called to those who were playing inside the house, and said:

"You must not make so much noise, or the Great Fire will come."

The children, who would not believe him, went on with their noisy play, and at last the Great Fire appeared. Little Kagssagssuk fled into the house, and cried:

"Lift me up. I must have my gloves, and they are up there!"

So they lifted him up to the drying frame under the roof.

And then they heard the Great Fire come hurrying into the house from without. He had a great live ribbon seal for a whip, and that whip had long claws. And then he began dragging the children out through the passage with his great whip, and each time he drew one out, that one was frizzled up. And at last there were no more. But before going away, the Great Fire reached up and touched with his finger a skin which was hanging on the drying frame.

As soon as the Great Fire had gone away, little Kagssagssuk crawled down from the drying frame and went over to the people who were gathered in the wizard's house, and told them what had happened. But none believed what he said.

"You have killed them yourself," they declared.

"Very well, then," he said, "if you think so, try to make a noise yourselves, like the children did."

And now they began cooking blubber above the entrance to the house, and when the oil was boiling and bubbling as hard as it could, they began making a mighty noise. And true enough, up came the Great Fire outside.

But little Kagssagssuk was not allowed to come into the house, and therefore he hid himself in the store shed. The Great Fire came into the house, and brought with it the live ribbon seal for a whip.

They heard it coming in through the passage, and then they poured boiling oil over it, and his whip being thus destroyed, the Great Fire went away.

But from that time onward, all the people of the village were unkind to little Kagssagssuk, and that although he had told the truth. Up to that time he had lived in the house of Umerdlugtoq, who was a great man, but now he was forced to stay outside always, and they would not let him come in. If he ventured to step in, though it were for no more than to dry his boots, Umerdlugtoq, that great man, would lift him up by the nostrils, and cast him over the high threshold again.

And little Kagssagssuk had two grandmothers; the one of these beat him as often as she could, even if he only lay out in the passage. But his other grandmother took pity on him, because he was the son of her daughter, who had been a woman like herself, and therefore she dried his clothes for him.

When, once in a while, that unfortunate boy did come in, Umerdlugtoq's folk would give him some tough walrus hide to eat, wishing only to give him something which they knew was too tough for him. And when they did so, he would take a little piece of stone and put it between his teeth, to help him, and when he had finished, put it back in his breeches, where he always kept it. When he was hungry, he would sometimes eat of the dogs' leavings on the ground outside, finding there walrus hide which even the dogs refused to eat.

He slept among the dogs, and warmed himself up on the roof, in the warm air from the smoke hole. But whenever Umerdlugtoq saw him warming himself there, he would haul him down by the nostrils.

Thus a long time passed, and it had been dark in the winter, and was beginning to grow light near the coming of spring. And now little Kagssagssuk began to go wandering about the country. Once when he was out, he met a big man, a giant, who was cutting up his catch, and on seeing him, Kagssagssuk cried out in a loud voice:

"Ho, you man there, give me a piece of that meat!"

But although he shouted as loudly as he could, that giant could not hear him. At last a little sound reached the big man's ears, and then he said:

"Bring me luck, bring me luck!"

And he threw down a little piece of meat on the ground, believing it was one of the dead who thus asked.

But little Kagssagssuk, who, young as he was, had already some helping spirits, made that little piece of meat to be a big piece, just as the dead can do, and ate as much as he could, and when he could eat no more, there was still so much left that he could hardly drag it away to hide it.

Some time after this, little Kagssagssuk said to his mother's mother:

"I have by chance become possessed of much meat,

and my thoughts will not leave it. I will therefore go out and look to it."

So he went off to the place where he had hidden it, and lo! it was not there. And he fell to weeping, and while he stood there weeping, the giant came up.

"What are you weeping for?"

"I cannot find the meat which I had hidden in a store-place here."

"Ho," said the giant, "I took that meat. I thought it had belonged to another one."

And then he said again: "Now let us play together." For he felt kindly towards that boy, and had pity on him.

And they two went off together. When they came to a big stone, the giant said: "Now let us push this stone." And they began pushing at the big stone until they twirled it round. At first, when little

Kagssagssuk tried, he simply fell backwards.

"Now once more. Make haste, make haste, once more. And there again, there is a bigger one."

And at last little Kagssagssuk ceased to fall over backwards, and was able instead to move the stones and twirl them round. And each time he tried with a larger stone than before, and when he had succeeded with that, a larger one still. And so he kept on. And at last he could make even the biggest stones twirl round in the air, and the stone said "leu-leu-leu-leu" in the air.

Then said the giant at last, seeing that they were equal in strength:

"Now you have become a strong man. But since it was by my fault that you lost that piece of meat, I will by magic means cause bears to come down to your village. Three bears there will be, and they will come right down to the village."

Then little Kagssagssuk went home, and having returned home, went up to warm himself as usual at the smoke hole. Then came the master of that house, as usual, and hauled him down by the nostrils. And afterwards, when he went to lie down among the dogs, his wicked grandmother beat him and them together, as was her custom. Altogether as if there were no strong man in the village at all.

But in the night, when all were asleep, he went down to one of the umiaks, which was frozen fast, and hauled it free.

Next morning when the men awoke, there was a great to-do.

"Haul. That umiak has been hauled out of the ice!"

"Haul. There must be a strong man among us!"

"Who can it be that is so strong?"

"Here is the mighty one, without a doubt," said Umerdlugtoq, pointing to little Kagssagssuk. But this he said only in mockery.

And a little time after this, the people about the village began to call out that three bears were in sight exactly as the giant had said. Kagssagssuk was inside, drying his boots. And while all the others were shouting eagerly about the place, he said humbly:

"If only I could borrow a pair of indoor boots from some one."

And at last, as he could get no others, he was obliged to take his grandmother's boots and put them on.

Then he went out, and ran off over the hard-trodden snow outside the houses, treading with such force that it seemed as if the footmarks were made in soft snow. And thus he went off to meet the bears.

"Haul. Look at Kagssagssuk. Did you ever see . . ."

"What is come to Kagssagssuk; what can it be?"

Umerdlugtoq was greatly excited, and so astonished that his eyes would not leave the boy. But little Kagssagssuk grasped the biggest of the bears, a mother with two half-grown cubs, grasped that bear with his naked fists, and wrung its neck, so that it fell down dead. Then he took those cubs by the back of the neck and hammered their skulls together until they too were dead.

Then little Kagssagssuk went back homeward with the biggest bear over his shoulders, and one cub under each arm, as if they had been no more than hares. Thus he brought them up to the house, and skinned them; then he set about building a fireplace large enough to put a man in. For he was now going to cook bears' meat for his grandmother, on a big flat stone.

Umerdlugtoq, that great man, now made haste to get away, taking his wives with him.

And Kagssagssuk took that old grandmother who was wont to beat him, and cast her on the fire, and she burned all up till only her stomach was left. His other grandmother was about to run away, but he held her back, and said:

"I shall now be kind to you, for you always used to dry my boots."

Now when Kagssagssuk had made a meal of the bears' meat, he set off in chase of those who had fled away. Umerdlugtoq had halted upon the top of a high hill, just on the edge of a precipice, and had pitched their tent close to the edge.

Up came Kagssagssuk behind him, caught him by the nostrils and held him out over the edge, and shook him so violently that his nostrils burst. And there stood Umerdlugtoq holding his nose. But Kagssagssuk said to him:

"Do not fear; I am not going to kill you. For you never used to kill me."

And then little Kagssagssuk went into the tent, and called out to him:

"Hi, come and look! I am in here with your wives!"

For in the old days, Umerdlugtoq had dared him even to look at them.

And having thus taken due vengeance, Kagssagssuk went back to his village, and took vengeance there on all those who had ever ill-treated him. And some time after, he went away to the southward, and lived with the people there.

It is also told that he got himself a kayak there, and went out hunting with the other men. But being so strong, he soon became filled with the desire to be feared, and began catching hold of children and crushing them. And therefore his fellow-villagers harpooned him one day when he was out in his kayak.

All this we have heard tell of Kagssagssuk.

QASIAGSSAQ, THE GREAT LIAR

QASIAGSSAQ, men say, was a great liar. His wife was called Qigdlugsuk. He could never sleep well at night, and being sleepless, he always woke his fellow-villagers when they were to go out hunting in the morning. But he never brought home anything himself.

One day when he had been out as usual in his kayak, without even sight of a seal, he said:

"It is no use my trying to be a hunter, for I never catch anything. I may as well make up some lie or other."

And at the same moment he noticed that one of his fellow-villagers was towing a big black seal over to an island, to land it there before going out for more. When that seal had been brought to land, Qasiagssaq rowed round behind the man, and stole it, and towed it back home.

His wife was looking out for him, going outside every now and then to look if he were in sight. And thus it was that coming out, she caught sight of a kayak coming in with something in tow. She shaded her eyes with both hands, one above the other, and looked through between them, gazing eagerly to try if she could make out who it was. The kayak with its seal in tow came rowing in, and she kept going out to look, and at last, when she came out as usual, she could see that it was really and truly Qasiagssaq, coming home with his catch in tow.

"Here is Qasiagssaq has made a catch," cried his fellow villagers. And when he came in, they saw that he had a great black seal in tow, with deep black markings all over the body. And the tow-line was thick with trappings of the finest narwhal tusk.

"Where did you get that tow-line?" they asked.

"I have had it a long time," he answered, " but have never used it before to-day."

After they had hauled the seal to land, his wife cut out the belly part, and when that was done, she shared out so much blubber and meat to the others that there was hardly anything left for themselves. And then she set about cooking a meal, with a shoulder-blade for a lamp, and another for a pot. And every time a kayak came in, they told the newcomer that Qasiagssaq had got a big black seal.

At last there was but one kayak still out, and when that one came in, they told him the same thing: "Qasiagssaq has actually got a big seal."

But this last man said when they told him:

"I got a big black seal to-day, and hauled it up on an island. But when I went back to fetch it, it was gone."

The others said again:

"The tow-line which Qasiagssaq was using to-day was furnished with toggles of pure narwhal tusk."

Later in the evening, Qasiagssaq heard a voice calling in at the window:

"You, Qasiagssaq, I have come to ask if you will give back that tow-line."

Qasiagssaq sprang up and said:

"Here it is; you may take it back now."

But his wife, who was beside him, said:

"When Qasiagssaq does such things, one cannot but feel shame for him."

"Hrrrr!" said Qasiagssaq to his wife, as if to frighten her. And after that he went about as if nothing had happened.

One day when he was out in his kayak as usual, he said:

"What is the use of my being out here, I who never

catch anything?"

And he rowed in towards land. When he reached the shore, he took off his breeches, and sat down on the ground, laying one knee across a stone. Then he took another stone to serve as a hammer, and with that he hammered both his knee-caps until they were altogether smashed.

And there he lay. He lay there for a long time, but at last he got up and went down to his kayak, and now he could only walk with little and painful steps. And when he came down to his kayak, he hammered and battered at that, until all the woodwork was broken to pieces. And then, getting into it, he piled up a lot of fragments of iceberg upon it, and even placed some inside his clothes, which were of raven's skin. And so he rowed home.

But all this while two women had been standing watching him. His wife was looking out for him as usual, shading her eyes with her hands, and when at last she caught sight of his kayak, and it came nearer, she could see that it was Qasiagssaq, rowing very slowly. And when then he reached the land, she said:

"What has happened to you now?"

"An iceberg calved."

And seeing her husband come home in such a case, his wife said to the others:

"An iceberg has calved right on top of Qasiagssaq, so that he barely escaped alive."

But when the women who had watched him came home, they said:

"We saw him to-day; he rowed in to land, and took off his breeches and hammered at his knee-caps with a stone; then he went down to his kayak and battered it to bits, and when that was done, he filled his kayak with ice, and even put ice inside his clothing."

But when his wife heard this, she said to him:

"When Qasiagssaq does such things, one cannot but feel shame for him."

"Hrrrr!" said Qasiagssaq, as if to frighten her. After that he lay still for a long while, waiting for his knees to heal, and when at last his knees were well again, he began once more to go out in his kayak, always without catching anything, as usual. And when he had thus been out one day as usual, without catching anything, he said to himself again:

"What is the use of my staying out here?"

And he rowed in to land. There he found a long stone, laid it on his kayak, and rowed out again. And when he came in sight of other kayaks that lay waiting for seal, he stopped still, took out his two small bladder floats made from the belly of a seal, tied the harpoon line to the stone in his kayak, and when that was done, he rowed away as fast as he could, while the kayaks that were waiting looked on. Then he disappeared from sight behind an iceberg, and when he came round on the other side, his bladder float was gone, and he himself was rowing as fast as he could towards land. His wife, who was looking out for him as usual, shading her eyes with her hands, said then:

"But what has happened to Qasiagssaq?"

As soon as a voice could reach the land, Qasiagssaq cried:

"Now you need not be afraid of breaking the handles of your knives; I have struck a great walrus, and it has gone down under water with my two small bladder floats. One or another of those who are out after seal will be sure to find it."

He himself remained altogether idle, and having come into his house, did not go out again. And as the kayaks began to come in, others went down to the shore and told them the news:

"Qasiagssaq has struck a walrus."

And this they said to all the kayaks as they came

home, but as usual, there was one of them that remained out a long time, and when at last he came back, late in the evening, they told him the same thing: "Qasiagssaq, it is said, has struck a walrus."

"That I do not believe, for here are his bladder floats; they had been tied to a stone, and the knot had worked loose."

Then they brought those bladder floats to Qasiagssaq and said:

"Here are your bladder floats; they were fastened to a stone, but the knot worked loose."

"When Qasiagssaq does such things, one cannot but feel shame for him," said his wife as usual.

"Hrrrr!" said Qasiagssaq, to frighten her.

And after that Qasiagssaq went about as if nothing had happened.

One day he was out in his kayak as usual at a place

where there was much ice; here he caught sight of a speckled seal, which had crawled up on to a piece of the ice. He rowed up to it, taking it unawares, and lifted his harpoon ready to throw, but just as he was about to throw, he looked at the point, and then he laid the harpoon down again, saying to himself: "Would it not be a pity, now, for that skin, which is to be used to make breeches for my wife, to be pierced with holes by the point of a harpoon?"

So he lay alongside the piece of ice, and began whistling to that seal.* And he was just about to grasp hold of it when the seal went down. But he watched it carefully, and when it came up again, he rowed over to it once more. Now he lifted his harpoon and was just about to throw, when again he caught sight of the point, and said to himself: "Would it not be a pity if that skin, which is to make breeches for my wife, should be pierced with holes by the point of a harpoon?" And again he cried out to try and frighten the seal, and down it went

**Speckled seal may often be caught in this fashion.*

again, and did not come up any more.

Once he heard that there lived an old couple in another village, who had lost their child. So Qasiagssaq went off there on a visit. He came to their place, and went into the house, and there sat the old couple mourning. Then he asked the others of the house in a low voice:

"What is the trouble here?"

"They are mourning," he was told.

"What for?" he asked.

"They have lost a child; their little daughter died the other day."

"What was her name?"

"Nipisartangivaq," they said.

Then Qasiagssaq cleared his throat and said in a loud voice:

"To-day my little daughter Nipisartangivaq is doubtless crying at her mother's side as usual."

Hardly had he said this when the mourners looked up eagerly, and cried:

"Ah, how grateful we are to you! *Now your little daughter can have all her things."

And they gave him beads, and the little girl's mother said:

"I have nothing to give you by way of thanks, but you shall have my cooking pot."

And when he was setting out again for home, they gave him great quantities of food to take home to his little girl. But when he came back to his own place, his fellow-villagers asked:

**The souls of the dead are supposed to be born again in the body of one named after them.*

"Wherever did you get all this?"

"An umiak started out on a journey, and the people in it were hurried and forgetful. Here are some things which they left behind them."

Towards evening a number of kayaks came in sight; it was people coming on a visit, and they had all brought meat with them. When they came in, they said:

"Tell Qasiagssaq and his wife to come down and fetch up this meat for their little girl."

"Qasiagssaq and his wife have no children; we know Qasiagssaq well, and his wife is childless."

When the strangers heard this, they would not even land at the place, but simply said:

"Then tell them to give us back the beads and the cooking pot."

And those things were brought, and given back to

them.

Then Qasiagssaq's wife said as usual:

"Now you have lied again. When you do such things, one cannot but feel shame for you."

"Hrrrr!" said Qasiagssaq, to frighten her, and went on as if nothing had happened.

Now it is said that Qasiagssaq's wife Qigdlugsuk had a mother who lived in another village, and had a son whose name was Ernilik. One day Qasiagssaq set out to visit them. He came to their place, and when he entered into the house, it was quite dark, because they had no blubber for their lamp, and the little child was crying, because it had nothing to eat. Qasiagssaq cleared his throat loudly and said:

"What is the matter with him?"

"He is hungry, as usual," said the mother.

Then said Qasiagssaq:

"How foolish I was not to take so much as a little blubber with me. Over in our village, seals are daily thrown away. You must come back with me to our place."

Next morning they set off together. When they reached the place, Qasiagssaq hurried up with the harpoon line in his hand, before his wife s mother had landed. And all she saw was that there was much carrion of ravens on Qasiagssaq's rubbish heap. Suddenly Qasiagssaq cried out:

"Ah! One of them has got away again!"

He had caught a raven in his snare. His wife cooked it, and their lamp was a shoulder-blade, and another shoulder-blade was their cooking pot, and when that meat was cooked, Qigdlugsuk's mother was given raven's meat to eat. Afterwards she was well fed by the other villagers there, and next morning when she was setting out to go home, they all gave her meat to take with her; all save Qasiagssaq, who gave her nothing.

And time went on, and once he was out as usual in his kayak, and when he came home in the evening, he said:

"I have found a dead whale; to-morrow we must all go out in the umiak and cut it up."

Next day many umiaks and kayaks set out to the eastward, and when they had rowed a long way in, they asked:

"Where is it?"

"Over there, beyond that little nest," he said.

And they rowed over there, and when they reached the place, there was nothing to be seen. So they asked again:

"Where is it?"

"Over there, beyond that little nest."

And they rowed over there, but when they reached the place, there was nothing to be seen. And again they asked:

"Where is it? Where is it?"

"Up there, beyond the little nest."

And again they reached the place and rowed round it, and there was nothing to be seen.

Then the others said:

"Qasiagssaq is lying as usual. Let us kill him."

But he answered:

"Wait a little; let us first make sure that it is a lie, and if you do not see it, you may kill me."

And again they asked:

"Where is it?"

"Yes . . . where was it now . . . over there beyond that little nest."

And now they had almost reached the base of that great fjord, and again they rounded a little ness farther in, and there was nothing to be seen. Therefore they said:

"He is only a trouble to us all: let us kill him."

And at last they did as they had said, and killed him.

THE EAGLE AND THE WHALE

IN a certain village there lived many brothers. And they had two sisters, both of an age to marry, and often urged them to take husbands, but they would not. At last one of the men said:

"What sort of a husband do you want, then? An eagle, perhaps? Very well, an eagle you shall have."

This he said to the one. And to the other he said:

"And you perhaps would like a whale? Well, a whale you shall have."

And then suddenly a great eagle came in sight, and it swooped down on the young girl and flew off with her to a high ledge of rock. And a whale also came in sight, and carried off the other sister, carrying her likewise to a ledge of rock.

After that the eagle and the girl lived together on a ledge of rock far up a high steep cliff. The eagle flew out over the sea to hunt, and while he was away, his wife would busy herself plaiting sinews for a line wherewith to lower herself down the rock. And while she was busied with that work, the eagle would sometimes appear, with a walrus in one claw and a narwhal in the other.

One day she tried the line, with which she was to lower herself down; it was too short. And so she plaited more.

But as time went on, the brothers began to long for their sister. And they all set to work making crossbows.

And there was in that village a little homeless boy, who was so small that he had not strength to draw a bow, but must get one of the others to draw it for him every time he wanted to shoot. When they had made all things ready, they went out to the place where their sister was, and called to her from the foot of the cliff, telling her to lower herself down. And this she did. As soon as her husband had gone out hunting, she lowered herself down and reached her brothers.

Towards evening, the eagle appeared out at sea, with a walrus in each claw, and as he passed the house of his wife's brothers, he dropped one down to them. But when he came home, his wife was gone. Then he simply threw his catch away, and flew, gliding on widespread wings, down to where those brothers were. But whenever the eagle tried to fly down to the house, they shot at it with their bows. And as none of them could hit, the little homeless boy cried:

"Let me try too!"

And then one of the others had to bend his bow for him. But when he shot off his arrow, it struck. And when then the eagle came fluttering down to earth, the others shot so many arrows at it that it could not quite touch the ground.

Thus they killed their sister's husband, who was a mighty hunter.

But the other sister and the whale lived together likewise. And the whale was very fond of her, and would hardly let her out of his sight for a moment.

But the girl here likewise began to feel homesick, and she also began plaiting a line of sinew threads, and her brothers, who were likewise beginning to long for their sister, set about making a swift-sailing umiak. And when they had finished it, and got it into the water, they said:

"Now let us see how fast it can go."

And then they got a guillemot which had its nest close by to fly beside them, while they tried to outdistance it by rowing. But when it flew past them, they cried:

"This will not do; the whale would overtake us at once. We must take this boat to pieces and build a new one." And so they took that boat to pieces and built a new one.

Then they put it in the water again and once more let the bird fly a race with them. And now the two kept side by side all the way, but when they neared the land, the bird was left behind.

One day the girl said as usual to the whale: "I must go outside a little."

"Stay here," said her husband, that great one.

"But I must go outside," said the girl.

Now he had a string tied to her, and this he would pull when he wanted her to come in again. And hardly had she got outside when he began pulling at the string.

"I am only just outside the passage," she cried. And then she tied the string by which she was held, to a stone, and ran away as fast as she could down hill, and the whale hauled at the stone, thinking it was his wife, and pulled it in. The brothers' house was just below the hillside where she was, and as soon as she came home, they fled away with her. But at the same moment, the whale came out from the passage way of its house, and rolled down into the sea. The umiak dashed off, but it seemed as if it were standing still, so swiftly did the whale overhaul it. And when the whale had nearly reached them, the brothers said to their sister:

"Throw out your hairband."

And hardly had she thrown it out when the sea foamed up, and the whale stopped. Then it went on after them again, and when it came up just behind the boat, the brothers said: "Throw out one of your mittens."

And she threw it out, and the sea foamed up, and the whale pounced down on it. And then she threw out the inner lining of one of her mittens, and then her outer frock and then her inner coat, and now they were close to land, but the whale was almost upon them. Then the brothers cried:

"Throw out your breeches!"

And at the same moment the sea was lashed into foam, but the umiak had reached the land. And the whale tried to follow, but was cast up on the shore as a white and sun-bleached bone of a whale.

THE TWO LITTLE OUTCASTS

THERE were two little boys and they had no father and no mother, and they went out every day hunting ptarmigan, and they had never any weapons save a bow. And when they had been out hunting ptarmigan, the men of that place were always very eager to take their catch.

One day they went out hunting ptarmigan as usual, but there were none. On their way, they came to some wild and difficult cliffs. And they looked down from that place into a ravine, and saw at the bottom a thing that looked like a stone. They went down towards it, and when they came nearer, it was a little house. And they went nearer still and came right to it. They climbed up on to the roof, and when they looked down through the air hole in the roof, they saw a little boy on the floor with a cutting-board for a kayak and a stick for a paddle. They called down to him, and he looked up, but then they hid themselves. When they looked down again, he was there as before, playing at being a man in a kayak. A second time they called to him, and then he ran to hide. And they went in then, and found him, sobbing a little, and pressing himself close in against the wall.

And they asked him:

"Do you live here all alone?"

And he answered: "No, my mother went out early this morning, and she is out now, as usual."

They said:

"We have come to be here with you because you are all alone."

And when they said this, he ventured to come out a little from the wall.

In the afternoon, the boy went out again and again and when he did so, they looked round the inside of the house, which was covered with fox skins, blue and white.

At last the boy came in, and said:

"Now I can see her, away to the south."

They looked out and saw her, and she seemed mightily big, having something on her back. And she came quickly nearer.

Then they heard a great noise, and that was the woman throwing down her burden. She came in hot and tired, and sat down, and said:

"Thanks, kind little boys. I had to leave him alone in the house, as usual, and now you have stayed with him while I was fearing for him on my way."

Then she turned to her son, and said:

"Have they not eaten yet?"

"No," said the boy. And when he had said that, she

went out, and came in with dried flesh of fox and reindeer, and a big piece of suet. And very glad they were to eat that food. At first they did not eat any of the dried fox meat, but when they tasted it, they found it was wonderfully good to eat.

Now when they had eaten their fill, they sat there feeling glad. And then the little boy whispered something in his mother's ear.

"He has a great desire for one of your sets of arrows, if you would not refuse to give it." And they gave him that.

In the evening, when they thought it was time to rest, a bed was made for them under the window, and when this was done the woman said:

"Now sleep, and do not fear any evil thing."

They slept and slept, and when they awoke, the woman had been awake a long time already.

And when they were setting off to go home again,

she paid them for their arrows with as much meat as they could carry; and when they went off, she said:

"Be sure you do not let any others come selling arrows."

But in the meantime, the people of the village had begun to fear for those two boys, because they did not come home. When at last they appeared in the evening, many went out to meet them. And it was a great load they had to carry.

"Where have you been?" they asked.

"We have been in a house with one who was not a real man."

They tasted the food they had brought. And it was wonderfully good to eat.

"That we were given in payment for one set of arrows," they said.

"We must certainly go out and sell arrows, too," said the others.

But the two told them: "No, you must not do that. For when we went away," she said: "Do not let any others come selling arrows."

But although this had been said to them, all fell to at once making arrows. And the next day they set out with the arrows on their backs. The two little boys did not desire to go, but went in despite of that, because the others ordered them.

Now when they came to the ravine, it looked as if that house were no longer there. And when they came down, not a stone of it was to be seen. They could not see so much as the two sheds or anything of them. And no one could now tell where that woman had gone.

And that was the last time they went out hunting ptarmigan.

ATDLARNEQ, THE GREAT GLUTTON

THIS is told of Atdlarneq: that he was a strong man, and if he rowed but a little way out in his kayak, he caught a seal. On no day did he fail to make a catch, and he was never content with only one.

But one day when he should have been out hunting seal, he only paddled along close to the shore, making towards the south. On the way he sighted a cape, and made towards it; and when he could see the sunny side, he spied a little house, quite near.

He thought: "I must wait until some one comes out."

And while he lay there, with his paddle touching the shore, a woman came out; she had a yellow band round her hair, and yellow seams to all her clothes.

Now he would have gone on shore, but he thought:

"I had better wait until another one comes out." And as he thought this, there came another woman out of the house. And like the first, she also had a yellow hair band, and yellow seams to all her clothes.

And he did not go on shore, but thought again: "I can wait for just one more."

And truly enough, there came yet another one, quite like the others. And like them also, she bore a dish in her hand. And now at last he went on shore and hauled up his kayak.

He went into the house, and they all received him very kindly. And they brought great quantities of food and set before him.

At last the evening came.

And now those three women began to go outside

again and again. And at last Atdlarneq asked:

"Why do you keep going out like that?"

When he asked them this, all answered at once:

"It is because we now expect our dear master home."

When he heard this, he was afraid, and hid himself behind the skin hangings. And he had hardly crawled in there when that master came home; Atdlarneq looked through a little hole, and saw him.

And his cheeks were made of copper.*

He had but just sat down, when he began to sniff, and said:

**There is a fabulous being in Inuit folklore supposed to have cheeks of copper, with which he can deliver terrible blows by a side movement of the head. Naughty children are frequently threatened with "Copper-cheeks" as a kind of bogey.*

"Hmm! There is a smell of people here."

And now Atdlarneq crawled out, seeing that the other had already smelt him. He had hardly shown himself, when the other asked very eagerly:

"Has he had nothing to eat yet?"
"No, he has not yet eaten."
"Then bring food at once."

And then they brought in a sack full of fish, and a big piece of blubber from the half of a black seal. And then the man said violently:

"You are to eat this all up, and if you do not eat it all up, I will thrash you with my copper cheeks!"

And now Atdlarneq began eagerly chewing blubber with his fish; he chewed and chewed, and at last he had eaten it all up. Then he went to the water bucket, and lifted it to his mouth and drank, and drank it all to the last drop.

Hardly had he done this when the man said:

"And now the frozen meat."

And they brought in the half of a black seal. And Atdlarneq ate and ate until there was no more left, save a very little piece.

When the man saw there was some not eaten, he cried out violently again:

"Give him some more to eat."

And when Atdlarneq had eaten again for a while, he did not wish to eat more. But then they brought in a whole black seal. And the man set that also before him, and cried:

"Eat that up too."

And so Atdlarneq was forced to stuff himself mightily once more.

He ate and ate, and at last he had eaten it all up. And again he emptied the water bucket.

After all that he felt very well indeed, and seemed hardly to have eaten until now. But that was because he had swallowed a little stalk of grass before he began.

So Atdlarneq slept, and next morning he went back home again. But after having thus nearly gorged himself to death, he never went southward again.

ANGANGUJUK

IT is said that Angangujuk's father was very strong. They had no other neighbours, but lived there three of them all alone. One day when the mother was going to scrape meat from a skin, she let the child play at kayak outside in the passage, near the entrance. And now and again she called to him: "Angangujuk!" And the child would answer from outside.

And once she called in this way, and called again, for there came no answer. And when no answer came again, she left the skin she was scraping, and began to search about. But she could not find the child. And now she began to feel greatly afraid, dreading her husband's return. And while she stood there feeling great fear of her husband, he came out from behind a rock, dragging a seal behind him.

Then he came forward and said:

"Where is our little son?"

"He vanished away from me this morning, after you had gone, when he was playing kayak-man out in the passage."

And when she had said this, her husband answered:

"It is you, wicked old hag, who have killed him. And now I will kill you."

To this his wife answered:

"Do not kill me yet, but wait a little, and first seek out one who can ask counsel of the spirits."

And now the husband began eagerly to search for such a one. He came home bringing wizards with him, and bade them try what they could do, and when they could not find the child, he let them go without giving them so much as a bite of meat.

And seeing that none of them could help him, he now sought for a very clever finder of hidden things, and meeting such a one at last, he took him home. Then he fastened a stick to his face, and made him lie down on the bedplace on his back.

And now he worked away with him until the spirit came. And when this had happened, the spirit finder declared:

"It would seem that spirits have here found a difficult task. He is up in a place between two great cliffs, and two old inland folk are looking after him."

Then they stopped calling spirits, and wandered away towards the east. They walked and walked, and at last they sighted a lot of houses. And when they came nearer, they saw the smoke coming out from all the smoke holes. It was the heat from inside coming out so. And the father looked in through a window, and saw that they were quarrelling about his child, and the child was crying.

"Who is to look after him?"

So he heard them saying inside the house; each one was eager to have the child. When the father saw this, he was very angry.

And the people inside asked the child:

"What would you like to eat?"

"No," said the child.

"Will you have seal meat?"

"No," said the child.

And there was nothing he cared to have. Therefore they asked him at last:

"Do you want to go home very much?"

Angangujuk answered quickly: "Yes." And his father was very greatly angered by now. And said to those with him:

"Try now to magic them to sleep."

And now the wizard began calling down a magic sleep upon those in the hut, and one by one they sank to sleep and began to snore. And fewer and fewer remained awake; at last there were only two. But then one of those two began to yawn, and at last rolled over and snored.

And now the great finder of hidden things began calling down sleep with all his might over that one remaining. And at last he too began to move towards the sleeping place. Then he began to yawn a little, and at last he also rolled over.

Now Angangujuk's father went in quickly, and now he caught up his son. But now the child had no clothes on. And looking for them, he saw them hung up on the drying frame. But the house was so high that they had to poke down the clothes with poles.

At last they came out, and walked and walked and came farther on. And it was now beginning to be light. As soon as they came to the place, they cut the moorings of the umiak, and hastily made all ready, and rowed out to the farthest islands. They had just moved away from land when they saw a number of people opposite the house.

But when the inland folk saw they had already moved out from the land, they went up to the house and beat it down, beating down roof and walls and all that there was of it.

After that time, Angangujuk's parents never again took up their dwelling on the mainland.

Here ends this story.

ATARSSUAQ

ATARSSUAQ had many enemies. But his many enemies tried in vain to hurt him, and they could not kill him. Then it happened that his wife bore him a son. Atarssuaq came back from his hunting one day, and found that he had a son. Then he took that son of his and bore him down to the water and threw him in. And waited until he began to kick out violently, and then took him up again. And so he did with him every day for long after, while the child was growing. And thus the boy became a very clever swimmer.

And one day Atarssuaq caught a fjord seal, and

took off the skin all in one piece, and dried it like a bladder, and made his son put it on when he went swimming.

One day he felt a wish to see how clever the boy had become. And said to him therefore:

"Go out now and swim, and I will follow after you."

And the father brought down his kayak and set it in the water, and his son watched him. And then he said:

"Now you swim out." And he made his father follow him out to sea, while he swam more and more under water. As soon as he came to the surface, his father rowed to where he was, but every time he took his throwing stick to cast a small harpoon, he disappeared.

And when his father thought they had done this long enough, he said:

"Now swim back to land, but keep under water as much as you can."

The son dived down, but it was a long time before he came up again. And now his father was greatly afraid. But at last the boy came up, a long way off. And then he rowed up to where he was, and laid one hand on his head, and said:

"Clever diver, clever diver, dear little clever one." And then he sniffed.

And a second time he said to him:

"Now swim under water a very long way this time."

So he dived down, and his father rowed forward all the time, to come to the place where he should rise, and feeling already afraid. His face moved as if he were beginning to cry, and he said:

"If only the sharks have not found him!" And he had just begun to cry when his son came up again. And then they went in to land, and the boy did not dive any more that day.

So clever had he now become.

And one day his father did not come back from his hunting. This was because of his enemies, who had killed him. Evening came, and next morning there was a kayak from the north. When it came in to the shore, the boy went down and said:

"To-morrow the many brothers will come to kill you all."

And the kayak turned at once and went back without coming on shore. Night passed and morning came. And in the morning when the boy awoke, he went to look out, and again, and many times. Once when he came out he saw many kayaks appearing from the northward. Then he went in and said to his mother:

"Now many kayaks are coming, to kill us all."

"Then put on your swimming dress," said his mother.

And he did so, and went down to the shore, and did not stop until he was quite close to the water. When the kayaks then saw him, they all rowed towards him, and said:

"He has fallen into the water."

When they came to the place where he had fallen in, they all began looking about for him, and while they were doing this, he came up just in front of the bone shoeing on the nose of one of the kayaks which lay quite away from the rest. When they spied him, each tried to outdo the others, and cried:

"Here he is!"

But then he dived down again. And this he continued to do. And in this manner he led all those kayaks out to the open sea, and when they had come a great way out, they sighted an iceberg which had run aground. When Atarssuaq's son came to this, he climbed up, by sticking his hands into the ice. And up above were two large pieces. And when he came close to the iceberg, he heard those in the kayaks saying among themselves:

"We can cut steps in the ice, and climb up to him."

And they began cutting steps in the iceberg, and at last the ice pick of the foremost came up over the edge. But now the boy took one of the great pieces of ice and threw it down upon them as they crawled up, so that it sent them all down again as it fell. And again he heard them say:

"It would be very foolish not to kill him. Let us climb up, and try to reach him this time."

And then they began crawling up one after another. But now the boy began as before, shifting the great piece of ice. And he waited until the head of the foremost one came up, and then he let it fall. And this time he also killed all those who had climbed on to the iceberg, after he had so lured them on to follow him.

But the others now turned back, and said:

"He will kill us all if we do not go."

And now the boy jumped down from the iceberg and swam to the kayaks and began tugging at their paddles, so that they turned over. But the men righted themselves again with their throwing sticks. And at last he was forced to hold them down himself under water till they drowned. And soon there were left no more of all those many kayaks, save only one. And when he looked closer, he saw that the man had no weapon but a stick for killing fish. And he rowed weeping in towards land, that man with no weapon but a stick. Then the boy pulled the paddle away from him, and he cried very much at that. Then he began paddling with his hands. But the boy gripped his hands from below, and then the man began crying furiously, and dared no longer put his hands in the water at all. And weeping very greatly he said:

"It is ill for me that ever I came out on this errand, for it is plain that I am to be killed."

The boy looked at him a little. And then said:

"You I will not kill. You may go home again." And he gave him back his paddle, and said to him as he was rowing away:

"Tell those of your place never to come out again thinking to kill us. For if they do not one of them will return alive."

Then Atarssuaq's son went home. And for some time he waited, thinking that more enemies might come. But none ever came against them after that time.

PUAGSSUAQ

THERE was once a wifeless man who always went out hunting ptarmigan. It became his custom always to go out hunting ptarmigan every day.

And when he was out one day, hunting ptarmigan as was his custom, he came to a place whence he could see out over a rocky valley. And it looked a good place to go. And he went there.

But before he had come to the bottom of the valley, he caught sight of something that looked like a stone. And when he could see quite clearly that it was not a stone at all, he went up to it. He walked and walked, and came to it at last.

Then he looked in, and saw an old couple sitting alone in there. And when he had seen this, he crawled very silently in through the passage way. And having come inside, he looked first a long time at them, and then he gave a little whistle. But nothing happened when he did so, and therefore he whistled a second time. And this time they heard the whistle, and the man nudged his wife and said:

"You, Puagssuaq, you can talk with the spirits. Take counsel with them now."

When he had said this, the wifeless man whistled again. And at this whistling, the man looked at his wife again and said earnestly:

"Listen! It sounds as if that might be the voice of a shore-dweller; one who catches miserable fish."

And now the wifeless man saw that the old one's wife was letting down her hair. And this was because she was now about to ask counsel of the spirits.

And he was now about to look at them again, when he saw that the passage way about him was beginning to close up. And it was already nearly closed up. But then it opened again of itself. Then the wifeless man thought only of coming out again from that place, and when the passage way again opened, he slipped out. And then he began running as fast as he could.

For a long time he ran on, with the thought that some one would surely come after him. But at last he came up the hillside, without having been pursued at all.

And when he came home, he told what had happened.

Here ends this story.

TUNGUJULUK AND SAUNIKOQ

TUNGUJULUK and Saunikoq were men from one village. And both were wizards. When they heard a spirit calling, one would change into a bear, and the other into a walrus.

Tungujuluk had a son, but Saunikoq had no children.

As soon as his son was old enough, Tungujuluk taught him to paddle a kayak. At this the other, Saunikoq, grew jealous, and began planning evil.

One morning when he awoke, he went out hunting seal as usual. He had been out some time, when he went up to an island, and called for his bearskin. When it came, he got into it, and moved off towards Tungujuluk's house. He landed a little way off, and then stole up to kill Tungujuluk's son. And when he came near, he saw him playing with the other children. But he did not know that his father had already come home, and was sitting busily at work on the kayak he was making for his son. He was just about to go up to them, when the boy went weeping home to his father, and when his father looked round, there was a big bear already close to them. He took a knife and ran towards it, and was just about to stab that bear, when it began to laugh. And then suddenly Tungujuluk remembered that his neighbour Saunikoq was able to take the shape of a bear. And he was now so angry that he had nearly stabbed him in spite of all, and it was a hard matter for him to hold back his knife.*

But he did not forget that happening. He waited until a long time had passed, and at last, many days later, when he awoke in the morning, he went out in his kayak. On the way he came to an island. And going up on to that island, he called his other shape to him. When it came, he crawled into it, and became a walrus. And when he had thus become a walrus, he went to that place where it was the custom for kayaks to hunt seal. And when he came near, he looked round, and sighted Saunikoq, who lay there waiting for seal.

Angiut, a "helping spirit," who knows all about everyone. Now he rose to the surface quite near him

**Flying race between two wizards, one of whom, unable to keep up, has fallen to earth, and is vainly begging the other to stop.*

and when Saunikoq saw him, he came over that way. And Saunikoq lifted his harpoon to throw it, and the stroke could not fail. Therefore he made himself small, and crept over to one side of the skin. And when he was struck, he floundered about a little, but not too violently, lest he should break the line. Then he swam away under water with the bladder float, and folded it up under his arm, and took out the air from it, and swam in towards land, and swam and swam until he came to the land near by where his kayak was lying. Then he went to it, and having taken out the point of the harpoon, he went out hunting.

He struck a black seal, and rowed home at once. And when he had come home, he said to his wife:

"Make haste and cook the breast piece."

And when that breast piece was cooked, and the other kayaks had come home, he made a meat feast, and Saunikoq, thinking nothing of any matter, came in with the others. When he came in, Tungujuluk made no sign of knowing anything, but went and took out the bladder and line from his kayak. And then all sat down to eat together. And they ate and were satisfied. And then each man began telling of his day's hunting.

At last Saunikoq said:

"To-day, when I struck a walrus, I did not think at all that it should cause me to lose my bladder float. Where that came up again is a thing we do not know. That bladder float of mine was lost."

And when Saunikoq had said this, Tungujuluk took that bladder and line and laid them beside the meat dish, and said:

"Whose can this bladder be, now, I wonder? Aha, at last I have paid you for the time when you came in the shape of a bear, and mocked us."

And when these words were said, the many who sat there laughed greatly. But Saunikoq got up and went away. And then next morning very early, he set out and rowed northward in his umiak. And since then he has not been seen. So great a shame did he feel.

ANARTEQ

THERE was once an old man, and he had only one son, and that son was called Anarteq. But he had many daughters. They were very fond of going out reindeer hunting to the eastward of their own place, in a fjord. And when they came right into the base of the fjord, Anarteq would let his sisters go up the hillside to drive the reindeer, and when they drove them so, those beasts came out into a big lake, where Anarteq could row out in his kayak and kill them all.

Thus in a few days they had their umiak filled with meat, and could go home again.

One day when they were out reindeer hunting, as

was their custom, and the reindeer had swum out, and Anarteq was striking them down, he saw a calf, and he caught hold of it by the tail and began to play with it. But suddenly the reindeer heaved up its body above the surface of the water, and kicked at the kayak so that it turned over. He tried to get up, but could not, because the kayak was full of water. And at last he crawled out of it.

The women looked at him from the shore, but they could not get out to help him, and at last they heard him say:

"Now the salmon are beginning to eat my belly."

And very slowly he went to the bottom.

Now when Anarteq woke again to his senses, he had become a salmon.

But his father was obliged to go back alone, and from that time, having no son, he must go out hunting as if he had been a young man. And he never again rowed up to those reindeer grounds where they had hunted before.

And now that Anarteq had thus become a salmon, he went with the others, in the spring, when the rivers break up, out into the sea to grow fat.

But his father, greatly wishing to go once more to their old hunting grounds, went there again as chief of a party, after many years had passed. His daughters rowed for him. And when they came in near to the base of the fjord, he thought of his son, and began to weep. But his son, coming up from the sea with the other salmon, saw the umiak, and his father in it, weeping. Then he swam to it, and caught hold of the paddle with which his father steered. His father was greatly frightened at this, and drew his paddle out of the water, and said:

"Anarteq had nearly pulled the paddle from my hand that time."

And for a long while he did not venture to put his paddle in the water again. When he did so at last, he saw that all his daughters were weeping. And a second time Anarteq swam quickly up to the umiak. Again the father tried to draw in his paddle when the son took hold of it, but this time he could not move it. But then at last he drew it quite slowly to the surface, in such a way that he drew his son up with it.

And then Anarteq became a man again, and hunted for many years to feed his kin.

THE GUILLEMOT THAT COULD TALK

A MAN from the south heard one day of a guillemot that could talk. It was said that this bird was to be found somewhere in the north, and therefore he set off to the northward. And toiled along north and north in an umiak.

He came to a village, and said to the people there:

"I am looking for a guillemot that can talk."

"Three days journey away you will find it."

Then he stayed there only that night, and went on again next morning. And when he came to a village, he had just asked his way, when one of the men there said:

"To-morrow I will go with you, and I will be a guide for you, because I know the way."

Next morning when they awoke, those two men set off together. They rowed and rowed and came in sight of a bird cliff. They came to the foot of that bird cliff, and when they stood at the foot and looked up, it was a mightily big bird cliff.

"Now where is that guillemot, I wonder?" said the man from the south. He had hardly spoken, when the man who was his guide said:

"Here, here is the nest of that guillemot bird."

And the man was prepared to be very careful when the bird came out of its nest. And it came out, that bird, and went to the side of the cliff and stared down at the kayaks, stretching its body to make it very long. And sitting up there, it said quite clearly:

"This, I think, must be that southern man, who has come far from a place in the south to hear a guillemot."

And the bird had hardly spoken, when he who was guide saw that the man from the south had fallen forward on his face. And when he lifted him up, that man was dead, having died of fright at hearing the bird speak.

Then seeing there was no other thing to be done, he covered up the body at the foot of the cliff below the guillemot's nest, and went home. And told the others of his place that he had covered him there below the guillemot's nest because he was dead. And the umiak and its crew of women stayed there, and wintered in that place.

Next summer, when they were making ready to go southward again, they had no man to go with them. But on the way that wifeless man procured food for them by catching fish, and when he had caught enough to fill a pot, he rowed in with his catch.

And in this way he led them southward. When they came to their own country, they had grown so fond of him that they would not let him go northward again. And so that wifeless man took a wife from among those women, because they would not let him go away to the north.

It is said that the skeleton of that wifeless man lies there in the south to this day.

KANAGSSUAQ

KANAGSSUAQ, men say, went out from his own place to live on a little island, and there took to wife the only sister of many brothers. And while he lived there with her, it happened once that the cold became so great that the sea between the islands was icebound, and they could no longer go out hunting. At last they had used up their store of food, and when that store of food was used up, and none of them could go out hunting, they all remained lying down from hunger and weakness.

Once, when there was open water to the south, where they often caught seal, Kanagssuaq took his kayak on his head and went out hunting. He rowed out in a northerly wind, with snow falling, and a heavy sea. And soon he came upon a number of black seal. He rowed towards them, to get within striking distance, but struck only a little fjord seal, which came up between him and the others. This one was easier to cut up, he said.

Now when he had got this seal, he took his kayak on his head again and went home across the ice. And his house-fellows shouted for joy when they saw the little creature he sent sliding in. Next day he went out again, and caught two black seal, and after that, he never went out without bringing home something.

The north wind continued, and the snow and the cold continued. When he lay out waiting for seal, as was now his custom, he often wished that he might meet with Kiliteraq, the great hunter from another place, who was the only one that would venture out in such weather. But this did not come about.

But now there was great dearth of food also in the place where Kiliteraq lived. And therefore Kiliteraq took his kayak on his head and went out across the ice to hunt seal. And coming some way, he sighted Kanagssuaq, who had already made his catch, and was just getting his tow-line out. As soon as he came up, Kanagssuaq cut away the whole of the belly skin and gave it to him. And Kiliteraq felt now a great desire for blubber, and took some good big pieces to chew.

And while he lay there, some black seal came up, and Kanagssuaq said:

"Row in to where they are."

And he rowed in to them and harpooned one, and killed it on the spot with that one stroke. He took his bladder float, to make a tow-line fast, and wound up the harpoon line, but before he had come to the middle, a breaking wave came rolling down on him. And it broke over him, and it seemed indeed as if there were no kayak there at all, so utterly was it hidden by that breaking wave. Then at last the bladder showed up behind the kayak, and a little after, the kayak itself came up, with the paddles held in a balancing position. Now for the second time he took his bladder and line, and just as he came to the place where the tow-line is made fast, there came another wave and washed over him so that he disappeared. And then he came up a second time, and as he came up, he said:

"I am now so far out that I cannot make my tow-line fast. Will you do this for me?"

And then Kanagssuaq made his tow-line fast, and as soon as he had taken the seal in tow, he rowed away in the thickly falling snow, and was soon lost to sight. When he came home, his many comrades in the village were filled with great thankfulness towards him. And thereafter it was as before; that he never came home without some catch.

A few days later, they awoke and saw that the snow was not falling near them now, but only far away on the horizon. And after that the weather became fine again. And when the spring came, they began hunting guillemots; driving them together in flocks and killing them so. This they did at that time.

And now one day they had sent their bird arrows showering down among the birds, and were busy placing the killed ones together in the kayaks. And then suddenly a kayak came in sight on the sunny side. And when that stranger came nearer, they looked eagerly to see who it might be. And when Kiliteraq came nearer for it was Kiliteraq who came he looked round among the kayaks, and when he saw that Kanagssuaq was among them, he thrust his way through and came, close up to him, and stuck his paddle in between the thongs on Kanagssuaq's kayak, and then loosened the skin over the opening of his own kayak, and put his hand in behind, and drew out a splendid tow-line made of walrus hide and beautifully worked with many beads of walrus tooth. And a second time he put in his hand, and took out now a piece of bearskin fashioned to the seat of a kayak. And these things he gave to Kanagssuaq, and said:

"Once in the spring, when I could not make my tow-line fast to a seal, you helped me, and made it fast. Here is that which shall thank you for that service."

And then he rowed away.

The sources of the legends are as follows:

POLAR ESKIMO, SMITH SOUND

SOUTH-EAST GREENLAND

WEST GREENLAND

GODTHAAB, WEST GREENLAND

SOUTH GREENLAND

UPERNIVIK, NORTH GREENLAND

Afterword

By W. W. Worster
excerpted from the original edition

The stories bring before us, more clearly, perhaps, than any objective study, the daily life of the Inuit, their habit of thought, their conception of the universe, and the curious "spirit world" which forms their primitive religion or mythology.

In point of form they are unique. The aim of the *Inuit* story teller is to pass the time during the long hours of darkness; if he can send his hearers to sleep, he achieves a triumph. Not infrequently a story-teller will introduce his chef-d'oeuvre with the proud declaration that "no one has ever heard this story to the end". The telling of the story thus becomes a kind of contest between his power of sustained invention and detailed embroidery on the one hand and his hearers' power of endurance on the other. Nevertheless, the stories are not as interminable as might be expected; we find also long and short variants of the same theme. In the present selection, versions of reasonable length have been preferred. The themes themselves are, of course, capable of almost infinite expansion.

The *Inuit* stories are magnificently heedless of much proportion. Any detail, whether of fact or fancy, can be expanded at will; a journey of many hundred miles may be summarized in a dozen words. It is thus a parallel in brief to the "Wandering" stories popular in Europe in the Middle Ages, when any kind of journey served as the string on which to gather all sorts of anecdote and adventure.

There seems to be a considerable intermingling of Christian culture and modern science in the general attitude towards life, but these foreign elements are coated over, as it were, like the speck of grit in an oyster, till they appear as concentrations of the native poetic spirit that forms their environment.

We find, too, constant evidence of derivation from the earliest, common sources of all folk-lore and myth; parallels to the fairy tales and legends of other lands and other ages.

Instances of friendship and affection between human beings and animals are found. Various resemblances to well-known fairy tales are discernible.

There is a general tendency towards anthropomorphic conception of supernatural beings. The life and domestic arrangements of "spirits" are mostly represented as very similar to those with which the story-teller and his hearers are familiar.

The style of narrative is peculiar. The stories open, as a rule, with some traditionally accepted gambit. The ending may occasionally point to a sort of moral, but the usual form is an equivalent. Some such hint is not infrequently necessary, since the "end" of a story often leaves considerable scope for further development.

It is a characteristic feature of these stories that one never knows what is going to happen. Poetic justice is often satisfied, but by no means always. One or two of them are naively weak and lacking in incident; we are constantly expecting something to happen, but nothing happens . . . still nothing happens . . . and the story ends. It is sometimes difficult to follow the exact course of a conversation or action between two personages.

The story-teller, while observing the traditional form, does not always do so uncritically. Occasionally he will throw in a little interpolation of his own, as if in apology.

Here and there, too, a touch of explanation may be inserted. And the manner in which this difference is viewed reveals two characteristic attitudes of mind, the blending of which is apparent throughout the *Inuit* culture of to-day. There is the attitude of condescension, the arrogant tolerance of the proselyte and the parvenu. At times, however, it is with precisely opposite view, mourning the present degeneration from earlier days.

And it is here, perhaps, that the stories reach their highest poetic level. This regret for the passing of "the former age" whether as an age of greater strength and virtue, or greater courage and skill, is a touching and most human trait. It gives to these poor *Inuit* hunters, far removed from the leisure and security that normally precede the growth of art, a place among the poets of the world.